The Atomic Nucleus

" The Positively Charged Core of an Atom "

Edited by Paul F. Kisak

Contents

Chapter 1

Atomic nucleus

The **nucleus** is the small, dense region consisting of protons and neutrons at the center of an atom. The atomic nucleus was discovered in 1911 by Ernest Rutherford based on the 1909 Geiger–Marsden gold foil experiment. After the discovery of the neutron in 1932, models for a nucleus composed of protons and neutrons were quickly developed by Dmitri Ivanenko[1] and Werner Heisenberg.[2][3][4][5][6] Almost all of the mass of an atom is located in the nucleus, with a very small contribution from the electron cloud. Protons and neutrons are bound together to form a nucleus by the nuclear force.

The diameter of the nucleus is in the range of 1.75 fm (1.75×10^{-15} m) for hydrogen (the diameter of a single proton)[7] to about 15 fm for the heaviest atoms, such as uranium. These dimensions are much smaller than the diameter of the atom itself (nucleus + electron cloud), by a factor of about 23,000 (uranium) to about 145,000 (hydrogen).

The branch of physics concerned with the study and understanding of the atomic nucleus, including its composition and the forces which bind it together, is called nuclear physics.

1.1   Introduction

1.1.1   History

Main article: Rutherford model

The nucleus was discovered in 1911, as a result of Ernest Rutherford's efforts to test Thomson's "plum pudding model" of the atom.[8] The electron had already been discovered earlier by J.J. Thomson himself. Knowing that atoms are neutral, Thomson postulated that there must be a positive charge as well. In his plum pudding model, Thomson stated that an atom consisted of negative electrons randomly scattered within a sphere of positive charge. Ernest Rutherford later devised an experiment, performed by Hans Geiger and Ernest Marsden under Rutherford's direction, that involved the deflection of alpha particles directed at a thin sheet of metal foil. He reasoned that if Thomson's model were correct, the positively charged alpha nuclei would easily pass through the foil with very little deviation in their paths as the foil should act in a manner as to be neutrally charged if the negative and positive charges are so intimately mixed as to make it appear neutral. To his surprise, many of the particles were deflected at very large angles. Because the mass of alpha particles is about 8000 times that of an electron, it became apparent that a very strong force must be present if it could deflect the massive and fast moving helium nuclei. He realized that the plum pudding model could not be accurate and that the deflections of the alpha particles could only be explained if the positive and negatives charges were in fact separated from each other and that the mass of the atom was a concentrated point of positive charge. Thus, the idea of a nuclear atom with a dense center of positive charge and mass became justified.

A model of the atomic nucleus showing it as a compact bundle of the two types of nucleons: protons (red) and neutrons (blue). In this diagram, protons and neutrons look like little balls stuck together, but an actual nucleus (as understood by modern nuclear physics) cannot be explained like this, but only by using quantum mechanics. In a nucleus which occupies a certain energy level (for example, the ground state), each nucleon can be said to occupy a range of locations.

1.1.2 Etymology

The term **nucleus** is from the Latin word *nucleus*, a diminutive of *nux* ("nut"), meaning the kernel (i.e., the "small nut") inside a watery type of fruit (like a peach). In 1844, Michael Faraday used the term to refer to the "central point of an atom". The modern atomic meaning was proposed by Ernest Rutherford in 1912.[9] The adoption of the term "nucleus" to atomic theory, however, was not immediate. In 1916, for example, Gilbert N. Lewis stated, in his famous article *The Atom and the Molecule*, that "the atom is composed of the *kernel* and an outer atom or *shell*"[10]

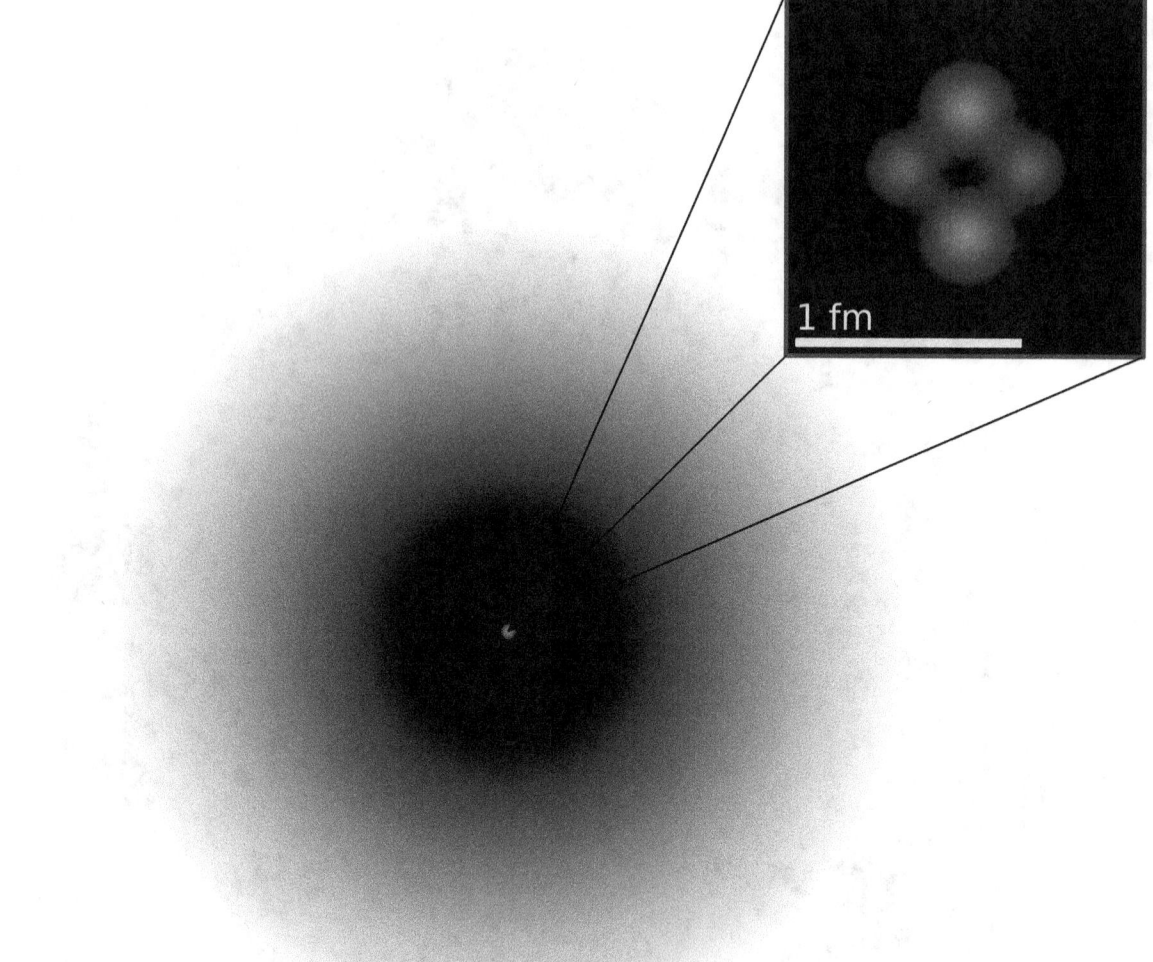

$$1 \text{ Å} = 100{,}000 \text{ fm}$$

*A figurative depiction of the helium−4 atom with the electron cloud in shades of gray. In the nucleus, the two protons and two neutrons are depicted in red and blue. This depiction shows the particles as separate, whereas in an actual helium atom, the protons are superimposed in space and most likely found at the very center of the nucleus, and the same is true of the two neutrons. Thus, all four particles are most likely found in exactly the same space, at the central point. Classical images of separate particles fail to model known charge distributions in very small nuclei. A more accurate image is that the spatial distribution of nucleons in helium's nucleus, although on a far smaller scale, is much closer to the helium **electron cloud** shown here, than to the fanciful nucleus image.*

1.1.3   Nuclear makeup

The nucleus of an atom consists of neutrons and protons, which in turn are the manifestation of fundamental particles, called quarks, that are held in association by the nuclear strong force in certain stable combinations of hadrons, called baryons. The nuclear strong force extends far enough from each baryon so as to bind the neutrons and protons together against the repulsive force of the positively charged protons. The nuclear strong force has a very short range and essentially drops to zero just beyond the edge of the nucleus. The collective action of the positively charged nucleus is to hold the electrically negative charged electrons in their orbits about the nucleus. The collection of negatively charged electrons orbiting the nucleus display an affinity for certain configurations and numbers of electrons that make their orbits stable. Which chemical element an atom represents is determined by the number of protons in the nucleus; the atom will have an equal number of electrons orbiting that nucleus. Individual chemical elements can create more stable electron configurations by combining to share their electrons. It is that sharing of electrons to create stable electronic orbits about the nucleus that appears to us as the chemistry of our macro world.

While protons define the entire charge of a nucleus and, hence, its chemical identity, neutrons are electrically neutral, but contribute to the mass of a nucleus to nearly the same extent as the protons. Neutrons explain the phenomenon of isotopes – varieties of the same chemical element which differ only in their atomic mass, not their chemical action.

1.2   Protons and neutrons

Protons and neutrons are fermions, with different values of the strong isospin quantum number, so two protons and two neutrons can share the same space wave function since they are not identical quantum entities. They sometimes are viewed as two different quantum states of the same particle, the *nucleon*.[11][12] Two fermions, such as two protons, or two neutrons, or a proton + neutron (the deuteron) can exhibit bosonic behavior when they become loosely bound in pairs.

In the rare case of a hypernucleus, a third baryon called a hyperon, with a different value of the strangeness quantum number can also share the wave function. However, the latter type of nuclei are extremely unstable and are not found on Earth except in high energy physics experiments.

The neutron has a positively charged core of radius ≈ 0.3 fm surrounded by a compensating negative charge of radius between 0.3 fm and 2 fm. The proton has an approximately exponentially decaying positive charge distribution with a mean square radius of about 0.8 fm.[13]

1.3   Forces

Nuclei are bound together by the residual strong force (nuclear force). The residual strong force is a minor residuum of the strong interaction which binds quarks together to form protons and neutrons. This force is much weaker *between* neutrons and protons because it is mostly neutralized within them, in the same way that electromagnetic forces *between* neutral atoms (such as van der Waals forces that act between two inert gas atoms) are much weaker than the electromagnetic forces that hold the parts of the atoms internally together (for example, the forces that hold the electrons in an inert gas atom bound to its nucleus).

The nuclear force is highly attractive at the distance of typical nucleon separation, and this overwhelms the repulsion between protons which is due to the electromagnetic force, thus allowing nuclei to exist. However, because the residual strong force has a limited range because it decays quickly with distance (see Yukawa potential), only nuclei smaller than a certain size can be completely stable. The largest known completely stable (e.g., stable to alpha, beta, and gamma decay) nucleus is lead-208 which contains a total of 208 nucleons (126 neutrons and 82 protons). Nuclei larger than this maximal size of 208 particles are unstable and (as a trend) become increasingly short-lived with larger size, as the number of neutrons and protons which compose them increases beyond this number. However, bismuth-209 is also stable to beta decay and has the longest half-life to alpha decay of any known isotope, estimated at a billion times longer than the age of the universe.

The residual strong force is effective over a very short range (usually only a few fermis; roughly one or two nucleon diameters) and causes an attraction between any pair of nucleons. For example, between protons and neutrons to form [NP] deuteron, and also between protons and protons, and neutrons and neutrons.

1.4   Halo nuclei and strong force range limits

The effective absolute limit of the range of the strong force is represented by halo nuclei such as lithium-11 or boron-14, in which dineutrons, or other collections of neutrons, orbit at distances of about ten fermis (roughly similar to the 8 fermi radius of the nucleus of uranium-238). These nuclei are not maximally dense. Halo nuclei form at the extreme edges of the chart of the nuclides—the neutron drip line and proton drip line—and are all unstable with short half-lives, measured in milliseconds; for example, lithium-11 has a half-life of 8.8 milliseconds.

Halos in effect represent an excited state with nucleons in an outer quantum shell which has unfilled energy levels "below" it (both in terms of radius and energy). The halo may be made of either neutrons [NN, NNN] or protons [PP, PPP]. Nuclei which have a single neutron halo include ^{11}Be and ^{19}C. A two-neutron halo is exhibited by ^{6}He, ^{11}Li, ^{17}B, ^{19}B and ^{22}C. Two-neutron halo nuclei break into three fragments, never two, and are called *Borromean nuclei* because of this behavior (referring to a system of three interlocked rings in which breaking any ring frees both of the others). ^{8}He and ^{14}Be both exhibit a four-neutron halo. Nuclei which have a proton halo include ^{8}B and ^{26}P. A two-proton halo is exhibited by ^{17}Ne and ^{27}S. Proton halos are expected to be more rare and unstable than the neutron examples, because of the repulsive electromagnetic forces of the excess proton(s).

1.5 Nuclear models

Although the standard model of physics is widely believed to completely describe the composition and behavior of the nucleus, generating predictions from theory is much more difficult than for most other areas of particle physics. This is due to two reasons:

- In principle, the physics within a nuclei can be derived entirely from quantum chromodynamics (QCD). In practice however, current computational and mathematical approaches for solving QCD in low-energy systems such as the nuclei are extremely limited. This is due to the phase transition that occurs between high-energy quark matter and low-energy hadronic matter, which renders perturbative techniques unusable, making it difficult to construct an accurate QCD-derived model of the forces between nucleons. Current approaches are limited to either phenomenological models such as the Argonne v18 potential or chiral effective field theory.[14]

- Even if the nuclear force is well constrained, a significant amount of computational power is required to accurately compute the properties of nuclei *ab initio*. Developments in many-body theory have made this possible for many low mass and relatively stable nuclei, but further improvements in both computational power and mathematical approaches are required before heavy nuclei or highly unstable nuclei can be tackled.

Historically, experiments have been compared to relatively crude models that are necessarily imperfect. None of these models can completely explain experimental data on nuclear structure.[15]

The nuclear radius (R) is considered to be one of the basic quantities that any model must predict. For stable nuclei (not halo nuclei or other unstable distorted nuclei) the nuclear radius is roughly proportional to the cube root of the mass number (A) of the nucleus, and particularly in nuclei containing many nucleons, as they arrange in more spherical configurations:

The stable nucleus has approximately a constant density and therefore the nuclear radius R can be approximated by the following formula,

$$R = r_0 A^{1/3}$$

where A = Atomic mass number (the number of protons Z, plus the number of neutrons N) and $r_0 = 1.25$ fm $= 1.25 \times 10^{-15}$ m. In this equation, the constant r_0 varies by 0.2 fm, depending on the nucleus in question, but this is less than 20% change from a constant.[16]

In other words, packing protons and neutrons in the nucleus gives *approximately* the same total size result as packing hard spheres of a constant size (like marbles) into a tight spherical or almost spherical bag (some stable nuclei are not quite spherical, but are known to be prolate).

1.5.1   Liquid drop model

Main article: Semi-empirical mass formula

Early models of the nucleus viewed the nucleus as a rotating liquid drop. In this model, the trade-off of long-range electromagnetic forces and relatively short-range nuclear forces, together cause behavior which resembled surface tension forces in liquid drops of different sizes. This formula is successful at explaining many important phenomena of nuclei, such as their changing amounts of binding energy as their size and composition changes (see semi-empirical mass formula), but it does not explain the special stability which occurs when nuclei have special "magic numbers" of protons or neutrons.

The terms in the semi-empirical mass formula, which can be used to approximate the binding energy of many nuclei, are considered as the sum of five types of energies (see below). Then the picture of a nucleus as a drop of incompressible liquid roughly accounts for the observed variation of binding energy of the nucleus:

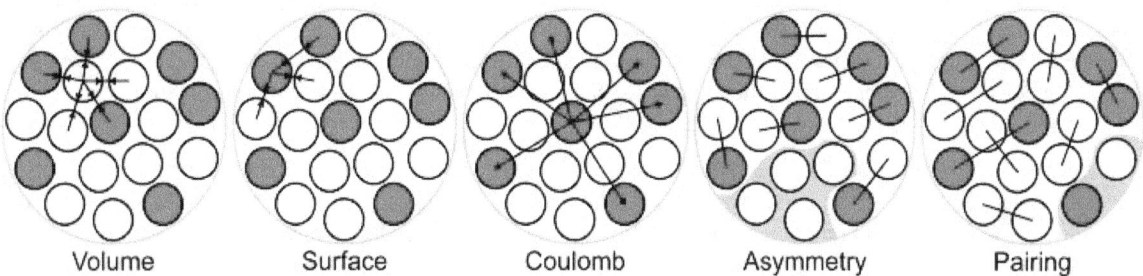

Volume   Surface   Coulomb   Asymmetry   Pairing

Volume energy. When an assembly of nucleons of the same size is packed together into the smallest volume, each interior nucleon has a certain number of other nucleons in contact with it. So, this nuclear energy is proportional to the volume.

Surface energy. A nucleon at the surface of a nucleus interacts with fewer other nucleons than one in the interior of the nucleus and hence its binding energy is less. This surface energy term takes that into account and is therefore negative and is proportional to the surface area.

Coulomb Energy. The electric repulsion between each pair of protons in a nucleus contributes toward decreasing its binding energy.

Asymmetry energy (also called Pauli Energy). An energy associated with the Pauli exclusion principle. Were it not for the Coulomb energy, the most stable form of nuclear matter would have the same number of neutrons as protons, since unequal numbers of neutrons and protons imply filling higher energy levels for one type of particle, while leaving lower energy levels vacant for the other type.

Pairing energy. An energy which is a correction term that arises from the tendency of proton pairs and neutron pairs to occur. An even number of particles is more stable than an odd number.

1.5.2   Shell models and other quantum models

Main article: Nuclear shell model

A number of models for the nucleus have also been proposed in which nucleons occupy orbitals, much like the atomic orbitals in atomic physics theory. These wave models imagine nucleons to be either sizeless point particles in potential wells, or else probability waves as in the "optical model", frictionlessly orbiting at high speed in potential wells.

In the above models, the nucleons may occupy orbitals in pairs, due to being fermions, which allows to explain even/odd Z and N effects well-known from experiments. The exact nature and capacity of nuclear shells differs from those of electrons in atomic orbitals, primarily because the potential well in which the nucleons move (especially in larger nuclei) is quite different from the central electromagnetic potential well which binds electrons in atoms. Some resemblance to atomic orbital models may be seen in a small atomic nucleus like that of helium-4, in which the two protons and two neutrons separately occupy 1s orbitals analogous to the 1s orbital for the two electrons in the helium atom, and achieve unusual stability for the same reason. Nuclei with 5 nucleons are all extremely unstable and short-lived, yet, helium-3, with 3 nucleons, is very stable even with lack of a closed 1s orbital shell. Another nucleus with 3 nucleons, the triton hydrogen-3 is unstable and will decay into helium-3 when isolated. Weak nuclear stability with 2 nucleons {NP} in the 1s orbital is found in the deuteron hydrogen-2, with only one nucleon in each of the proton and neutron potential wells. While each nucleon is a fermion, the {NP} deuteron is a boson and thus does not follow Pauli Exclusion for close packing within shells. Lithium-6 with 6 nucleons is highly stable without a closed second 1p shell orbital. For light nuclei with total nucleon numbers 1 to 6 only those with 5 do not show some evidence of stability. Observations of beta-stability of light nuclei outside closed shells indicate that nuclear stability is much more complex than simple closure of shell orbitals with magic numbers of protons and neutrons.

For larger nuclei, the shells occupied by nucleons begin to differ significantly from electron shells, but nevertheless, present nuclear theory does predict the magic numbers of filled nuclear shells for both protons and neutrons. The closure of the stable shells predicts unusually stable configurations, analogous to the noble group of nearly-inert gases in chemistry. An example is the stability of the closed shell of 50 protons, which allows tin to have 10 stable isotopes, more than any other element. Similarly, the distance from shell-closure explains the unusual instability of isotopes which have far from stable numbers of these particles, such as the radioactive elements 43 (technetium) and 61 (promethium), each of which is preceded and followed by 17 or more stable elements.

There are however problems with the shell model when an attempt is made to account for nuclear properties well away from closed shells. This has led to complex *post hoc* distortions of the shape of the potential well to fit experimental data, but the question remains whether these mathematical manipulations actually correspond to the spatial deformations in real nuclei. Problems with the shell model have led some to propose realistic two-body and three-body nuclear force effects involving nucleon clusters and then build the nucleus on this basis. Two such cluster models are the Close-Packed Spheron Model of Linus Pauling and the 2D Ising Model of MacGregor.[15]

1.5.3   Consistency between models

Main article: Nuclear structure

As with the case of superfluid liquid helium, atomic nuclei are an example of a state in which both (1) "ordinary" particle physical rules for volume and (2) non-intuitive quantum mechanical rules for a wave-like nature apply. In superfluid helium, the helium atoms have volume, and essentially "touch" each other, yet at the same time exhibit strange bulk properties, consistent with a Bose–Einstein condensation. The latter reveals that they also have a wave-like nature and do not exhibit standard fluid properties, such as friction. For nuclei made of hadrons which are fermions, the same type of condensation does not occur, yet nevertheless, many nuclear properties can only be explained similarly by a combination of properties of particles with volume, in addition to the frictionless motion characteristic of the wave-like behavior of objects trapped in Erwin Schrödinger's quantum orbitals.

1.6   See also

- Giant resonance

- List of particles

- Nuclear medicine

- Radioactivity

- Semi-empirical mass formula

1.7   References

[1] Iwanenko, D.D., The neutron hypothesis, Nature **129** (1932) 798.

[2] Heisenberg, W. (1932). "Über den Bau der Atomkerne. I". *Z. Phys.* **77**: 1–11. Bibcode:1932ZPhy...77....1H. doi:10.

.[3]Heisenberg,W. (1932). "Über den Bau der Atomkerne.II".*Z.Phys.* **78**(3–4):156–164.Bibcode:1932ZPhy...78..156H. doi:10.1007/BF01337585.

[4] Heisenberg, W. (1933). "Über den Bau der Atomkerne. III". *Z. Phys.* **80** (9–10): 587–596. Bibcode:1933ZPhy...80..587H. doi:10.1007/BF01335696.

[5] Miller A. I. *Early Quantum Electrodynamics: A Sourcebook*, Cambridge University Press, Cambridge, 1995, ISBN 0521568919, pp. 84–88.

[6] Bernard Fernandez and Georges Ripka (2012). "Nuclear Theory After the Discovery of the Neutron". *Unravelling the Mystery of the Atomic Nucleus: A Sixty Year Journey 1896 — 1956*. Springer. p. 263. ISBN 9781461441809. Retrieved 15 February 2013.

[7] Geoff Brumfiel (July 7, 2010). "The proton shrinks in size". *Nature.* doi:10.1038/news.2010.337.

[8] *Rutgers University.* "The Rutherford Experiment". physics.rutgers.edu. Retrieved February 26, 2013.

[9] D. Harper. "Nucleus". *Online Etymology Dictionary.* Retrieved 2010-03-06.

[10]G.N.Lewis(1916)."The Atom and the Molecule".*Journal of the American Chemical Society***38**(4):4.doi:10.1021/ja02261a.

[11] A.G. Sitenko, V.K. Tartakovskiĭ (1997). *Theory of Nucleus: Nuclear Structure and Nuclear Interaction.* Kluwer Academic. p. 3. ISBN 0-7923-4423-5.

[12] M.A. Srednicki (2007). *Quantum Field Theory.* Cambridge University Press. pp. 522–523. ISBN 978-0-521-86449-7.

[13] J.-L. Basdevant, J. Rich, M. Spiro (2005). *Fundamentals in Nuclear Physics.* Springer. p. 155. ISBN 0-387-01672-4.

[14]Machleidt,R.;Entem,D.R. (2011). "Chiral effectivefield theory and nuclear forces".*Physics Reports***503**(1):1–75.arXiv:1105. Bibcode:2011PhR...503....1M. doi:10.1016/j.physrep.2011.02.001.

[15] N.D. Cook (2010). *Models of the Atomic Nucleus* (2nd ed.). Springer. p. 57 ff. ISBN 978-3-642-14736-4.

[16] K.S. Krane (1987). *Introductory Nuclear Physics.* Wiley-VCH. ISBN 0-471-80553-X.

1.8   External links

- The Nucleus – a chapter from an online textbook

- The LIVEChart of Nuclides – IAEA in Java or HTML

- Article on the"nuclear shell model,"giving nuclear shellfilling for the various elements.Accessed.

- Timeline: Subatomic Concepts, Nuclear Science & Technology.

Chapter 2

Radioactive decay

For particle decay in a more general context, see Particle decay. For more information on hazards of various kinds of radiation from decay, see Ionizing radiation.

"Radioactive" redirects here. For other uses, see Radioactive (disambiguation).

"Radioactivity" redirects here. For other uses, see Radioactivity (disambiguation).

Radioactive decay, also known as **nuclear decay** or **radioactivity**, is the process by which a nucleus of an unstable atom

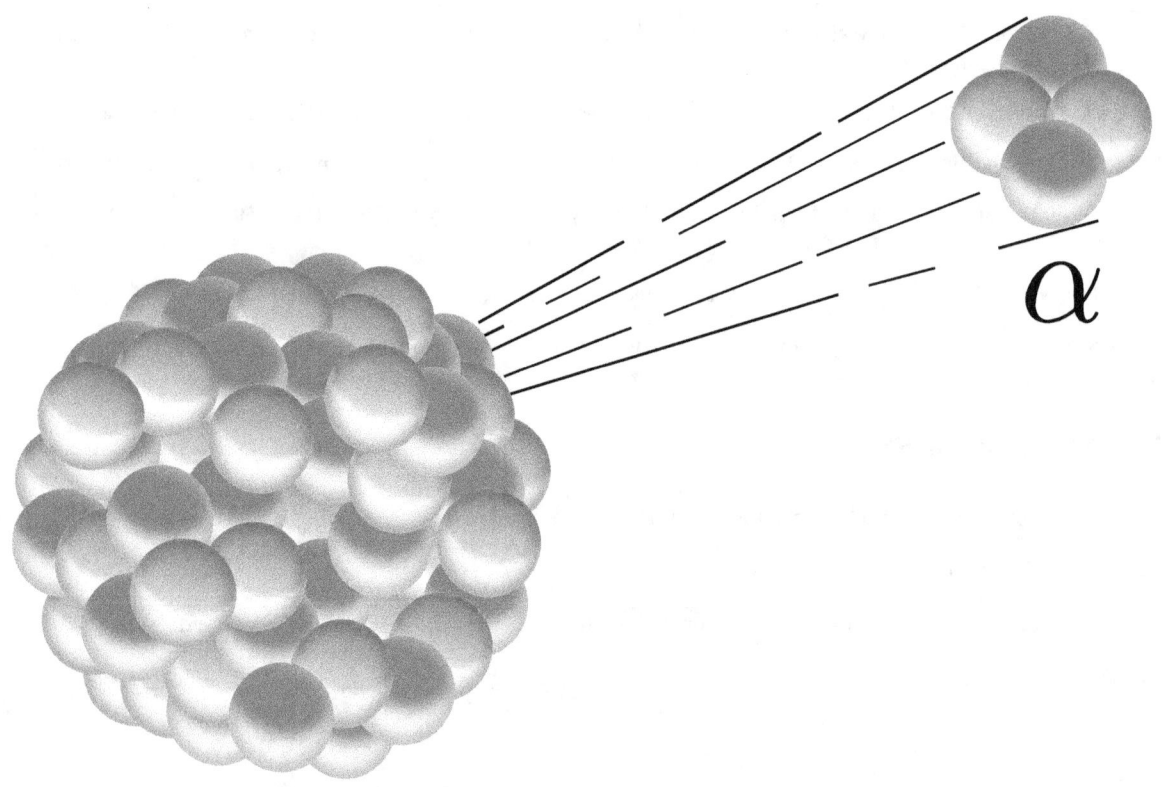

Alpha decay is one type of radioactive decay, in which an atomic nucleus emits an alpha particle, and thereby transforms (or "decays") into an atom with a mass number decreased by 4 and atomic number decreased by 2.

loses energy by emitting radiation. A material that spontaneously emits such radiation — which includes alpha particles, beta particles, gamma rays and conversion electrons — is considered **radioactive**.

Radioactive decay is a stochastic (i.e. random) process at the level of single atoms, in that, according to quantum theory, it is impossible to predict when a particular atom will decay.[1][2][3][4] The chance that a given atom will decay never changes, that is, it does not matter how long the atom has existed. For a large collection of atoms however, the decay rate for that collection can be calculated from their measured decay constants or half-lives. This is the basis of radiometric dating. The half-lives of radioactive atoms have no known limits for shortness or length of duration, and range over 55 orders of magnitude in time.

There are many types of radioactive decay (see table below). A decay, or loss of energy, results when an atom with one type of nucleus, called the *parent radionuclide* (or *parent radioisotope*[note 1]), transforms into an atom with a nucleus in a different state, or with a nucleus containing a different number of protons and neutrons. The product is called the *daughter nuclide*. In some decays, the parent and the daughter nuclides are different chemical elements, and thus the decay process results in the creation of an atom of a different element. This is known as a nuclear transmutation.

The first decay processes to be discovered were alpha decay, beta decay, and gamma decay. Alpha decay occurs when the nucleus ejects an alpha particle (helium nucleus). This is the most common process of emitting nucleons, but in rarer types of decays, nuclei can eject protons, or in the case of cluster decay specific nuclei of other elements. Beta decay occurs when the nucleus emits an electron or positron and a neutrino, in a process that changes a proton to a neutron or the other way about. The nucleus may capture an orbiting electron, causing a proton to convert into a neutron in a process called electron capture. All of these processes result in a nuclear transmutation.

By contrast, there are radioactive decay processes that do not result in a nuclear transmutation. The energy of an excited nucleus may be emitted as a gamma ray in a process called gamma decay, or be used to eject an orbital electron by its interaction with the excited nucleus, in a process called internal conversion. Highly excited neutron-rich nuclei, formed as the product of other types of decay, occasionally lose energy by way of neutron emission, resulting in a change of an element from one isotope to another. Another type of radioactive decay results in products that are not defined, but appear in a range of "pieces" of the original nucleus. This decay, called spontaneous fission, happens when a large unstable nucleus spontaneously splits into two (and occasionally three) smaller daughter nuclei, and generally leads to the emission of gamma rays, neutrons, or other particles from those products.

For a summary table showing the number of stable and radioactive nuclides in each category, see radionuclide. There exist twenty-nine chemical elements on Earth that are radioactive. They are those that contain thirty-four radionuclides that date before the time of formation of the solar system. Well-known examples are uranium and thorium, but also included are naturally occurring long-lived radioisotopes such as potassium-40. Another fifty or so shorter-lived radionuclides, such as radium and radon, found on Earth, are the products of decay chains that began with the primordial nuclides, and ongoing cosmogenic processes, such as the production of carbon-14 from nitrogen-14 by cosmic rays. Radionuclides may also be produced artificially in particle accelerators or nuclear reactors, resulting in 650 of these with half-lives of over an hour, and several thousand more with even shorter half-lives. See this list of nuclides for a list by half life.

2.1   History of discovery

Radioactivity was discovered in 1896 by the French scientist Henri Becquerel, while working with phosphorescent materials.[5] These materials glow in the dark after exposure to light, and he suspected that the glow produced in cathode ray tubes by X-rays might be associated with phosphorescence. He wrapped a photographic plate in black paper and placed various phosphorescent salts on it. All results were negative until he used uranium salts. The uranium salts caused a blackening of the plate in spite of the plate being wrapped in black paper. These radiations were given the name "Becquerel Rays".

It soon became clear that the blackening of the plate had nothing to do with phosphorescence, as the blackening was also produced by non-phosphorescent salts of uranium and metallic uranium. It became clear from these experiments that there was a form of invisible radiation that could pass through paper and was causing the plate to react as if exposed to light.

At first, it seemed as though the new radiation was similar to the then recently discovered X-rays. Further research by Becquerel, Ernest Rutherford, Paul Villard, Pierre Curie, Marie Curie, and others showed that this form of radioactivity was significantly more complicated. Rutherford was the first to realize that all such elements decay in accordance with the same mathematical exponential formula. Rutherford and his student Frederick Soddy were the first to realize that many decay processes resulted in the transmutation of one element to another. Subsequently, the radioactive displacement law

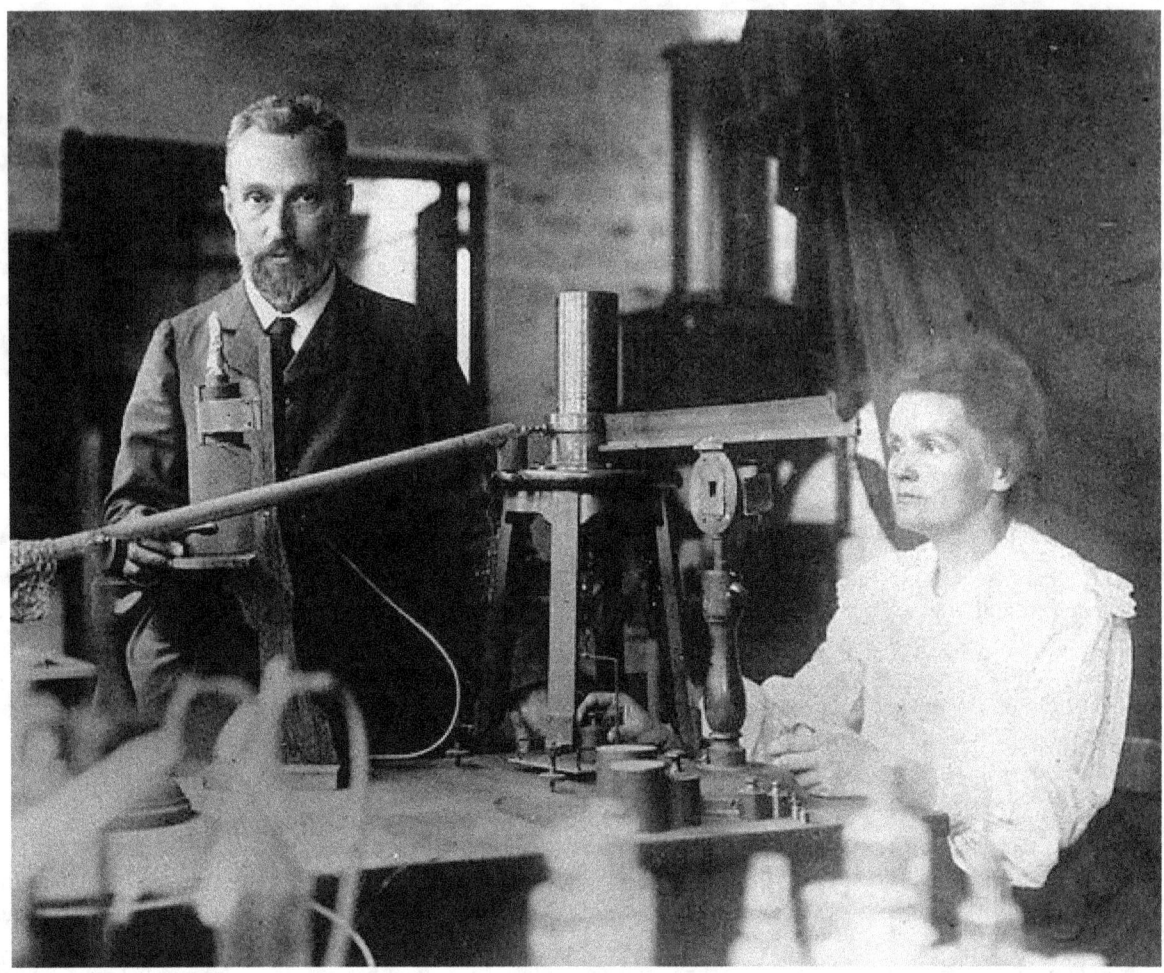

Pierre and Marie Curie in their Paris laboratory, before 1907

of Fajans and Soddy was formulated to describe the products of alpha and beta decay.[6][7].

The early researchers also discovered that many other chemical elements, besides uranium, have radioactive isotopes. A systematic search for the total radioactivity in uranium ores also guided Pierre and Marie Curie to isolate two new elements: polonium and radium. Except for the radioactivity of radium, the chemical similarity of radium to barium made these two elements difficult to distinguish.

2.2 Early health dangers

The dangers of ionizing radiation due to radioactivity and X-rays were not immediately recognized.

2.2.1 X-rays

The discovery of x-rays by Wilhelm Röntgen in 1895 led to widespread experimentation by scientists, physicians, and inventors. Many people began recounting stories of burns, hair loss and worse in technical journals as early as 1896. In February of that year, Professor Daniel and Dr. Dudley of Vanderbilt University performed an experiment involving X-raying Dudley's head that resulted in his hair loss. A report by Dr. H.D. Hawks, of his suffering severe hand and chest burns in an X-ray demonstration, was the first of many other reports in *Electrical Review*.[8]

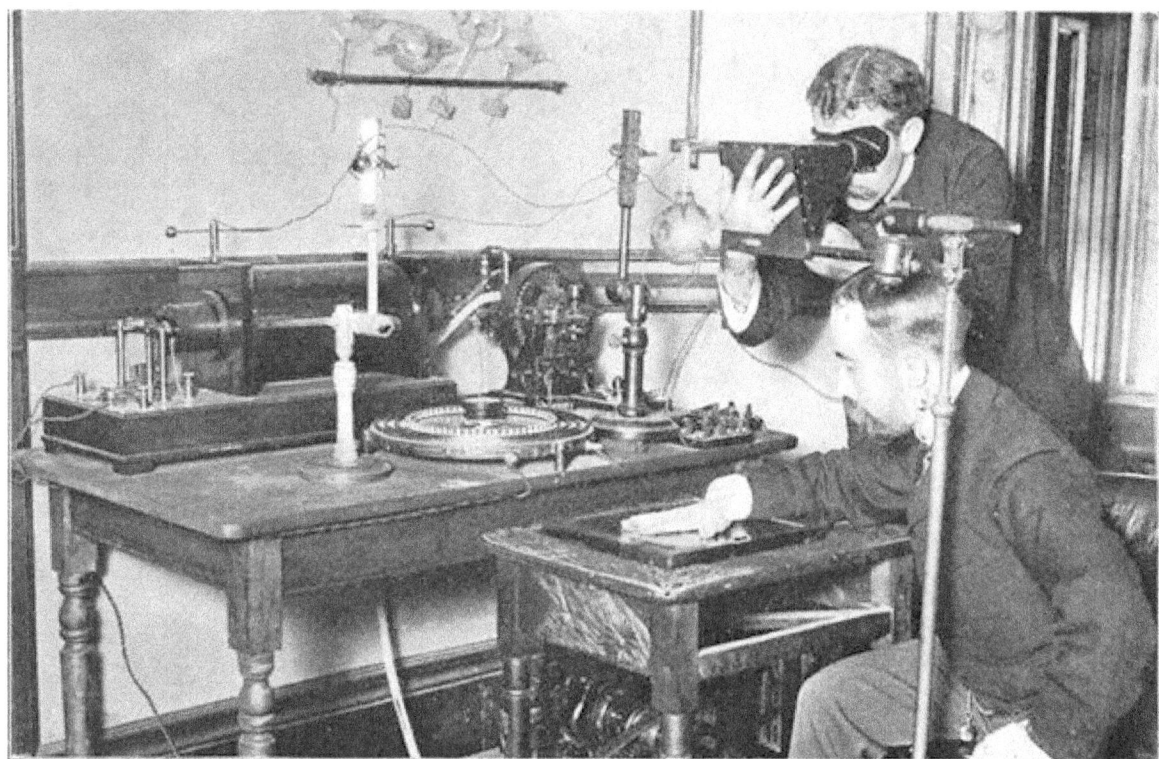

Taking an X-ray image with early Crookes tube apparatus in 1896. The Crookes tube is visible in the centre. The standing man is viewing his hand with a fluoroscope screen; this was a common way of setting up the tube. No precautions against radiation exposure are being taken; its hazards were not known at the time.

Other experimenters including Elihu Thomson, and Nikola Tesla also reported burns. Thomson deliberately exposed a finger to an X-ray tube over a period of time and suffered pain, swelling, and blistering.[9] Other effects, including ultraviolet rays and ozone were sometimes blamed for the damage,[10] and many physicians still claimed that there were no effects from X-ray exposure at all.[9]

Despite this, there were some early systematic hazard investigations, and as early as 1902 William Herbert Rollins wrote almost despairingly that his warnings about the dangers involved in careless use of X-rays was not being heeded, either by industry or by his colleagues. By this time Rollins had proved that X-rays could kill experimental animals, could cause a pregnant guinea pig to abort, and that they could kill a fetus.[11] He also stressed that "animals vary in susceptibility to the external action of X-light" and warned that these differences be considered when patients were treated by means of X-rays.

2.2.2 Radioactive substances

However, the biological effects of radiation due to radioactive substances were less easy to gauge. This gave the opportunity for many physicians and corporations to market radioactive substances as patent medicines. Examples were radium enema treatments, and radium-containing waters to be drunk as tonics. Marie Curie protested against this sort of treatment, warning that the effects of radiation on the human body were not well understood. Curie later died from aplastic anaemia, likely caused by exposure to ionizing radiation. By the 1930s, after a number of cases of bone necrosis and death of radium treatment enthusiasts, radium-containing medicinal products had been largely removed from the market (radioactive quackery).

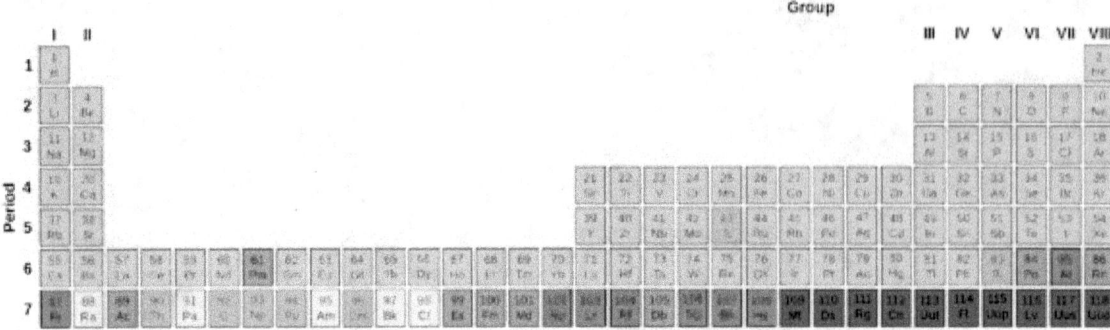

Radioactivity is characteristic of elements with large atomic number. Elements with at least one stable isotope are shown in light blue. Green shows elements whose most stable isotope has a half-life measured in millions of years. Yellow and orange are progressively more unstable, with half-lives in thousands or hundreds of years, down toward one day. Red and purple show highly and extremely radioactive elements where the most stable isotopes exhibit half-lives measured on the order of one day and much less.

2.2.3 Radiation protection

Main article: Radiation protection
See also: Sievert and Ionizing radiation

Only a year after Röntgen's discovery of X rays, the American engineer Wolfram Fuchs (1896) gave what is probably the first protection advice, but it was not until 1925 that the first International Congress of Radiology (ICR) was held and considered establishing international protection standards. The effects of radiation on genes, including the effect of cancer risk, were recognized much later. In 1927, Hermann Joseph Muller published research showing genetic effects and, in 1946, was awarded the Nobel prize for his findings.

The second ICR was held in Stockholm in 1928 and proposed the adoption of the rontgen unit, and the 'International X-ray and Radium Protection Committee' (IXRPC) was formed. Rolf Sievert was named Chairman, but a driving force was George Kaye of the British National Physical Laboratory. The committee met in 1931, 1934 and 1937.

After World War II the increased range and quantity of radioactive substances being handled as a result of military and civil nuclear programmes led to large groups of occupational workers and the public being potentially exposed to harmful levels of ionising radiation. This was considered at the first post-war ICR convened in London in 1950, when the present International Commission on Radiological Protection (ICRP) was born.[12] Since then the ICRP has developed the present international system of radiation protection, covering all aspects of radiation hazard.

2.3 Units of radioactivity

The International System of Units (SI) unit of radioactive activity is the becquerel (Bq), named in honour of the scientist Henri Becquerel. One Bq is defined as one transformation (or decay or disintegration) per second.

An older unit of radioactivity is the curie, Ci, which was originally defined as "the quantity or mass of radium emanation in equilibrium with one gram of radium (element)".[13] Today, the curie is defined as 3.7×10^{10} disintegrations per second, so that 1 curie (Ci) = 3.7×10^{10} Bq. For radiological protection purposes, although the United States Nuclear Regulatory Commission permits the use of the unit curie alongside SI units,[14] the European Union European units of measurement directives required that its use for "public health ... purposes" be phased out by 31 December 1985.[15]

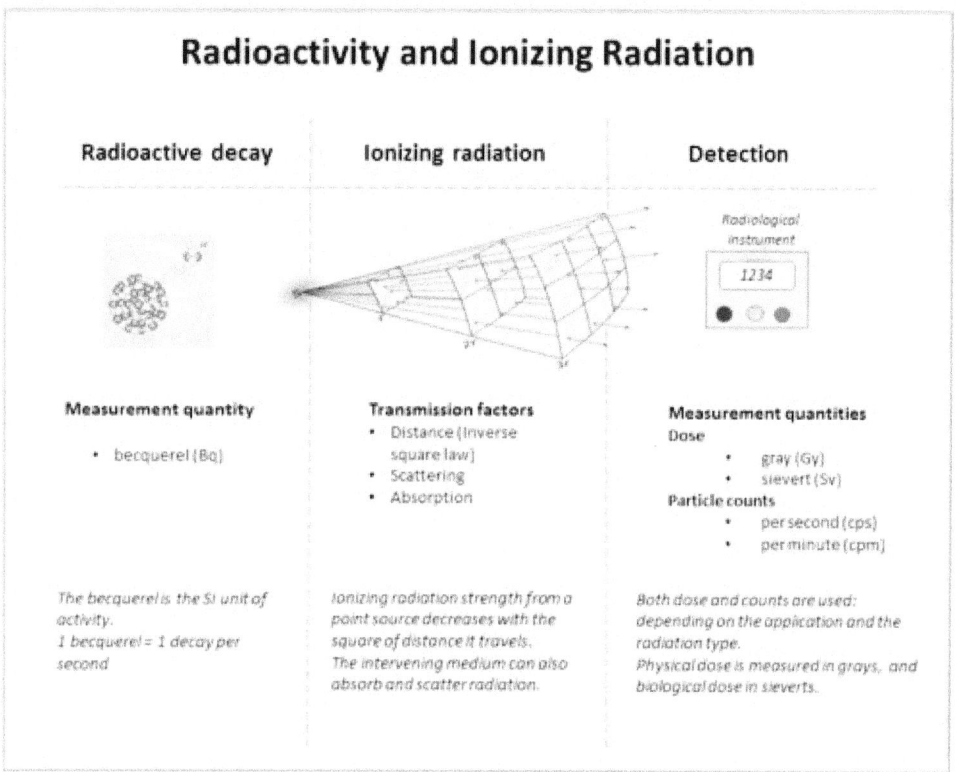

Graphic showing relationships between radioactivity and detected ionizing radiation

2.4   Types of decay

Early researchers found that an electric or magnetic field could split radioactive emissions into three types of beams. The rays were given the names alpha, beta, and gamma, in order of their ability to penetrate matter. While alpha decay was seen only in heavier elements of atomic number 52 (tellurium) and greater, the other two types of decay were produced by all of the elements. Lead, atomic number 82, is the heaviest element to have any isotopes stable (to the limit of measurement) to radioactive decay. Radioactive decay is seen in all isotopes of all elements of atomic number 83 (bismuth) or greater. Bismuth, however, is only very slightly radioactive.

In analysing the nature of the decay products, it was obvious from the direction of the electromagnetic forces applied to the radiations by external magnetic and electric fields that alpha particles carried a positive charge, beta particles carried a negative charge, and gamma rays were neutral. From the magnitude of deflection, it was clear that alpha particles were much more massive than beta particles. Passing alpha particles through a very thin glass window and trapping them in a discharge tube allowed researchers to study the emission spectrum of the captured particles, and ultimately proved that alpha particles are helium nuclei. Other experiments showed beta radiation, resulting from decay and cathode rays, were high-speed electrons. Likewise, gamma radiation and X-rays were found to be high-energy electromagnetic radiation.

The relationship between the types of decays also began to be examined: For example, gamma decay was almost always found to be associated with other types of decay, and occurred at about the same time, or afterwards. Gamma decay as a separate phenomenon, with its own half-life (now termed isomeric transition), was found in natural radioactivity to be a result of the gamma decay of excited metastable nuclear isomers, which were in turn created from other types of decay.

Although alpha, beta, and gamma radiations were most commonly found, other types of emission were eventually discovered. Shortly after the discovery of the positron in cosmic ray products, it was realized that the same process that

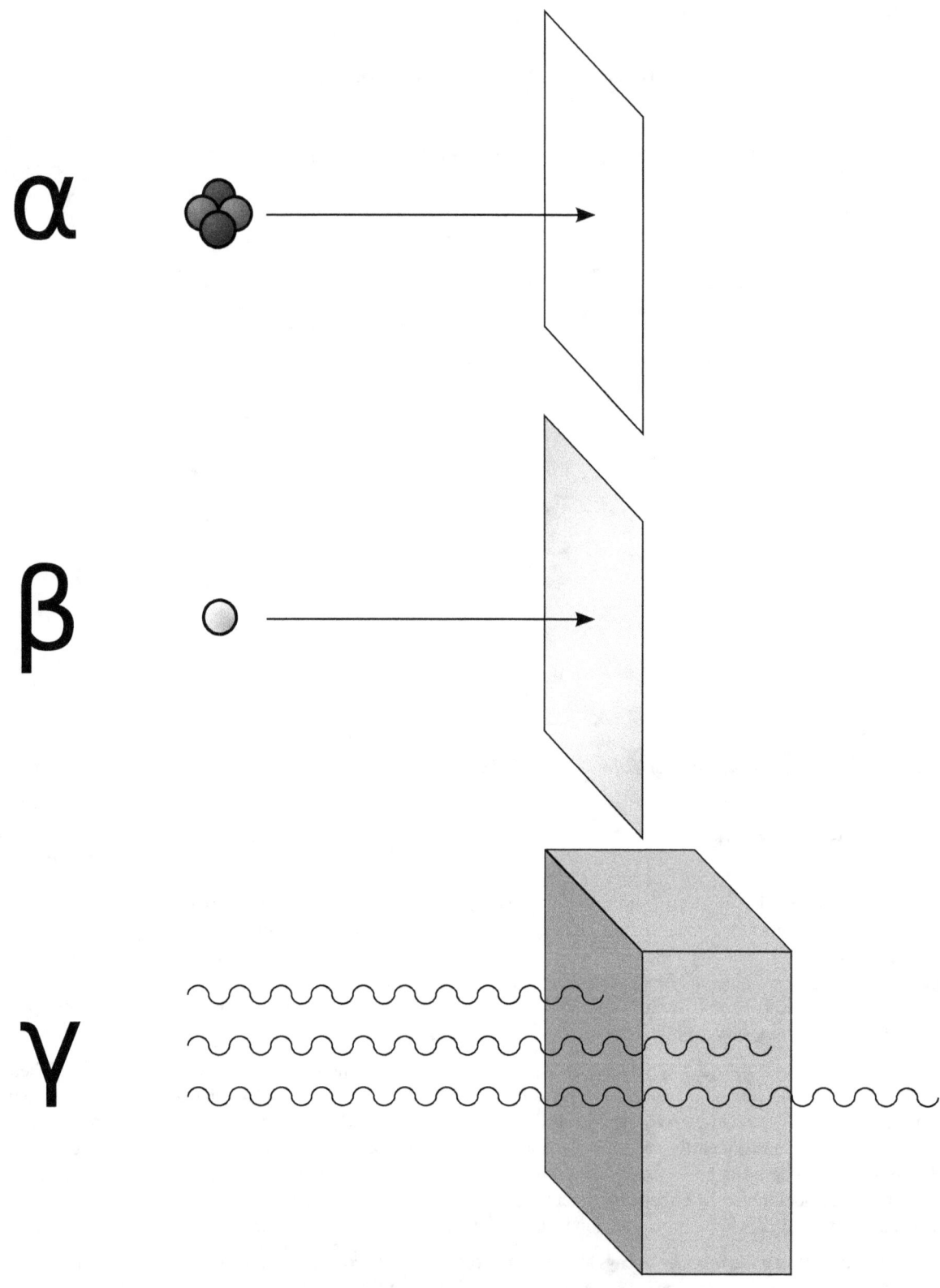

Alpha particles may be completely stopped by a sheet of paper, beta particles by aluminium shielding. Gamma rays can only be reduced by much more substantial mass, such as a very thick layer of lead.

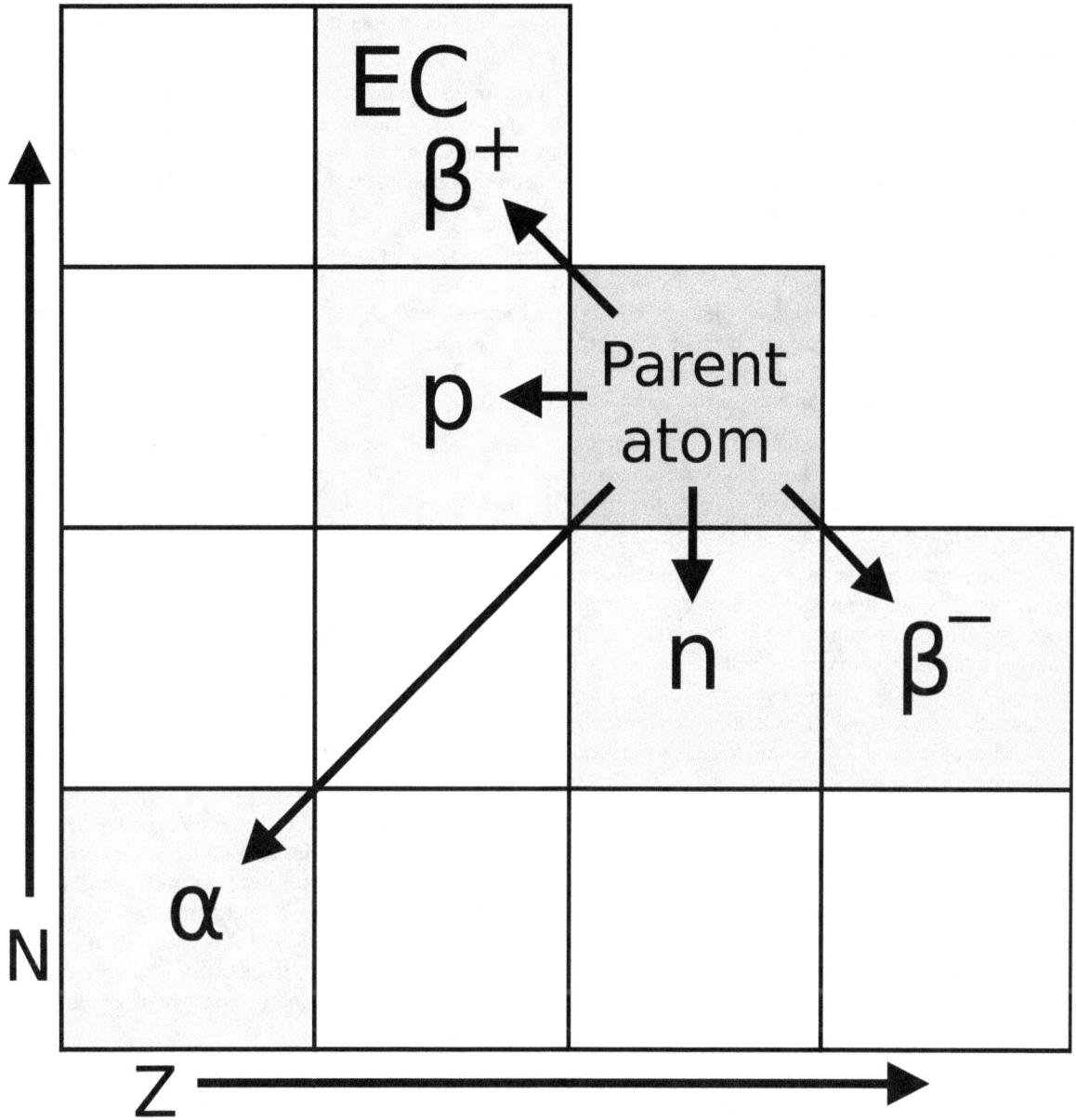

Transition diagram for decay modes of a radionuclide, with neutron number N and atomic number Z (shown are α, β±, p+, and n⁰ emissions, EC denotes electron capture).

operates in classical beta decay can also produce positrons (positron emission). In an analogous process, instead of emitting positrons and neutrinos, some proton-rich nuclides were found to capture their own atomic electrons, a process called electron capture, and subsequently emit only a neutrino and usually also a gamma ray. Each of these types of decay involves the capture or emission of nuclear electrons or positrons, and acts to move a nucleus toward the ratio of neutrons to protons that has the least energy for a given total number of nucleons, consequently producing a more stable nucleus.

A theoretical process of positron capture, analogous to electron capture, is possible in antimatter atoms, but has not been observed as antimatter atoms are rarely available.[16] This would require antimatter atoms at least as complex as beryllium-7, which is the lightest known isotope of normal matter to undergo decay by electron capture.

Shortly after the discovery of the neutron in 1932, Enrico Fermi realized that certain rare beta-decay reactions immediately yield neutrons as a decay particle (neutron emission). Isolated proton emission was eventually observed in some elements. It was also found that some heavy elements may undergo spontaneous fission into products that vary in composition. In a phenomenon called cluster decay, specific combinations of neutrons and protons other than alpha particles (helium nuclei) were found to be spontaneously emitted from atoms.

Other types of radioactive decay of different mechanisms were found to emit previously-seen particles. An example is internal conversion, which results in electron and sometimes high-energy photon emission, although it involves neither beta nor gamma decay. A neutrino is not emitted, and neither the electron nor photon originate in the nucleus. Internal conversion decay, like isomeric transition gamma decay and neutron emission, involves the release of energy by an excited nuclide, without the transmutation of one element into another.

Rare events that involve a combination of two beta-decay type events happening simultaneously are known (see below). Any decay process that does not violate the conservation of energy or momentum laws (and perhaps other particle conservation laws) is permitted to happen, although not all have been detected. An interesting example discussed in a final section, is bound state beta decay of rhenium-187. In this process, an inverse of electron capture, beta electron-decay of the parent nuclide is not accompanied by beta electron emission, because the beta particle has been captured into the K-shell of the emitting atom. An antineutrino, however, is emitted.

Radionuclides can undergo a number of different reactions. These are summarized in the following table. A nucleus with mass number A and atomic number Z is represented as (A, Z). The column "Daughter nucleus" indicates the difference between the new nucleus and the original nucleus. Thus, $(A - 1, Z)$ means that the mass number is one less than before, but the atomic number is the same as before.

If energy circumstances are favorable, a given radionuclide may undergo many competing types of decay, with some atoms decaying by one route, and others decaying by another. An example is copper-64, which has 29 protons, and 35 neutrons, which decays with a half-life of about 12.7 hours. This isotope has one unpaired proton and one unpaired neutron, so either the proton or the neutron can decay to the opposite particle. This particular nuclide (though not all nuclides in this situation) is almost equally likely to decay through proton decay, producing a positron emission (18%), or through electron capture (43%), as it does through neutron decay by electron emission (39%). The excited energy states resulting from these decays which fail to end in a ground energy state, also produce later internal conversion and gamma decay in almost 0.5% of the time.

Radioactive decay results in a reduction of summed rest mass, once the released energy (the *disintegration energy*) has escaped in some way. Although decay energy is sometimes defined as associated with the difference between the mass of the parent nuclide products and the mass of the decay products, this is true only of rest mass measurements, where some energy has been removed from the product system. This is true because the decay energy must always carry mass with it, wherever it appears (see mass in special relativity) according to the formula $E = mc^2$. The decay energy is initially released as the energy of emitted photons plus the kinetic energy of massive emitted particles (that is, particles that have rest mass). If these particles come to thermal equilibrium with their surroundings and photons are absorbed, then the decay energy is transformed to thermal energy, which retains its mass.

Decay energy therefore remains associated with a certain measure of mass of the decay system, called invariant mass, which does not change during the decay, even though the energy of decay is distributed among decay particles. The energy of photons, the kinetic energy of emitted particles, and, later, the thermal energy of the surrounding matter, all contribute to the invariant mass of the system. Thus, while the sum of the rest masses of the particles is not conserved in radioactive decay, the *system* mass and system invariant mass (and also the system total energy) is conserved throughout any decay process. This is a restatement of the equivalent laws of conservation of energy and conservation of mass.

2.5 Radioactive decay rates

The *decay rate*, or *activity*, of a radioactive substance is characterized by:

Constant quantities:

- The *half-life*—$t_{1/2}$, is the time taken for the activity of a given amount of a radioactive substance to decay to half of its initial value; see List of nuclides.

- The *decay constant*— λ, "lambda" the inverse of the mean lifetime, sometimes referred to as simply *decay rate*.

- The *mean lifetime*— τ, "tau" the average lifetime of a radioactive particle before decay.

Although these are constants, they are associated with the statistical behavior of populations of atoms. In consequence, predictions using these constants are less accurate for minuscule samples of atoms.

In principle a half-life, a third-life, or even a $(1/\sqrt{2})$-life, can be used in exactly the same way as half-life; but the mean life and half-life $t_{1/2}$ have been adopted as standard times associated with exponential decay.

Time-variable quantities:

- *Total activity*— A, is the number of decays per unit time of a radioactive sample.

- *Number of particles*—N, is the total number of particles in the sample.

- *Specific activity*—SA, number of decays per unit time per amount of substance of the sample at time set to zero ($t = 0$). "Amount of substance" can be the mass, volume or moles of the initial sample.

These are related as follows:

$$t_{1/2} = \frac{\ln(2)}{\lambda} = \tau \ln(2)$$

$$A = -\frac{dN}{dt} = \lambda N$$

$$S_A a_0 = -\frac{dN}{dt}\bigg|_{t=0} = \lambda N_0$$

where N_0 is the initial amount of active substance — substance that has the same percentage of unstable particles as when the substance was formed.

2.6   Mathematics of radioactive decay

For the mathematical details of exponential decay in general context, see exponential decay.
For related derivations with some further details, see half-life.
For the analogous mathematics in 1st order chemical reactions, see Consecutive reactions.

2.6.1   Universal law of radioactive decay

Radioactivity is one very frequently given example of exponential decay. The law describes the statistical behaviour of a large number of nuclides, rather than individual atoms. In the following formalism, the number of nuclides or the nuclide population N, is of course a discrete variable (a natural number)—but for any physical sample N is so large that it can be treated as a continuous variable. Differential calculus is needed to set up differential equations for the modelling the behaviour of the nuclear decay.

The mathematics of radioactive decay depend on a key assumption that a nucleus of a radionuclide has no "memory" or way of translating its history into its present behavior. A nucleus does not "age" with the passage of time. Thus, the probability of its breaking down does not increase with time, but stays constant no matter how long the nucleus has existed. This constant probability may vary greatly between different types of nuclei, leading to the many different observed decay rates. However, whatever the probability is, it does not change. This is in marked contrast to complex objects which do show aging, such as automobiles and humans. These systems do have a chance of breakdown per unit of time, that increases from the moment they begin their existence.

One-decay process

Consider the case of a nuclide A that decays into another B by some process $A \rightarrow B$ (emission of other particles, like electron neutrinos ν
e and electrons e$^-$ as in beta decay, are irrelevant in what follows). The decay of an unstable nucleus is entirely random and it is impossible to predict when a particular atom will decay.[1] However, it is equally likely to decay at any instant in time. Therefore, given a sample of a particular radioisotope, the number of decay events $-dN$ expected to occur in a small interval of time dt is proportional to the number of atoms present N, that is[17]

$$-\frac{dN}{dt} \propto N.$$

Particular radionuclides decay at different rates, so each has its own decay constant λ. The expected decay $-dN/N$ is proportional to an increment of time, dt:

The negative sign indicates that N decreases as time increases, as the decay events follow one after another. The solution to this first-order differential equation is the function:

$$N(t) = N_0\, e^{-\lambda t} = N_0\, e^{-t/\tau},$$

where N_0 is the value of N at time $t = 0$.[17]

We have for all time t:

$$N_A + N_B = N_{\text{total}} = N_{A0},$$

where N_{total} is the constant number of particles throughout the decay process, which is equal to the initial number of A nuclides since this is the initial substance.

If the number of non-decayed A nuclei is:

$$N_A = N_{A0}e^{-\lambda t}$$

then the number of nuclei of B, i.e. the number of decayed A nuclei, is

$$N_B = N_{A0} - N_A = N_{A0} - N_{A0}e^{-\lambda t} = N_{A0}\left(1 - e^{-\lambda t}\right).$$

The number of decays observed over a given interval obeys Poisson statistics. If the average number of decays is <N>, the probability of a given number of decays N is[17]

$$P(N) = \frac{\langle N \rangle^N \exp(-\langle N \rangle)}{N!}.$$

Chain-decay processes

Chain of two decays

Now consider the case of a chain of two decays: one nuclide A decaying into another B by one process, then B decaying into another C by a second process, i.e. $A \rightarrow B \rightarrow C$. The previous equation cannot be applied to the decay chain, but can be generalized as follows. Since A decays into B, *then* B decays into C, the activity of A adds to the total number of B nuclides in the present sample, *before* those B nuclides decay and reduce the number of nuclides leading to the later sample. In other words, the number of second generation nuclei B increases as a result of the first generation nuclei decay of A, and decreases as a result of its own decay into the third generation nuclei C.[18] The sum of these two terms gives the law for a decay chain for two nuclides:

$$\frac{dN_B}{dt} = -\lambda_B N_B + \lambda_A N_A.$$

The rate of change of N_B, that is dN_B/dt, is related to the changes in the amounts of A and B, N_B can increase as B is produced from A and decrease as B produces C.

Re-writing using the previous results:

The subscripts simply refer to the respective nuclides, i.e. N_A is the number of nuclides of type A, N_{A0} is the initial number of nuclides of type A, λ_A is the decay constant for A - and similarly for nuclide B. Solving this equation for N_B gives:

$$N_B = \frac{N_{A0}\lambda_A}{\lambda_B - \lambda_A} \left(e^{-\lambda_A t} - e^{-\lambda_B t}\right).$$

In the case where B is a stable nuclide ($\lambda_B = 0$), this equation reduces to the previous solution:

$$\lim_{\lambda_B \to 0} \left[\frac{N_{A0}\lambda_A}{\lambda_B - \lambda_A} \left(e^{-\lambda_A t} - e^{-\lambda_B t}\right)\right] = \frac{N_{A0}\lambda_A}{0 - \lambda_A} \left(e^{-\lambda_A t} - 1\right) = N_{A0}\left(1 - e^{-\lambda_A t}\right),$$

as shown above for one decay. The solution can be found by the integration factor method, where the integrating factor is $e^{\lambda_B t}$. This case is perhaps the most useful, since it can derive both the one-decay equation (above) and the equation for multi-decay chains (below) more directly.

Chain of any number of decays

For the general case of any number of consecutive decays in a decay chain, i.e. $A_1 \rightarrow A_2 \cdots \rightarrow A_i \cdots \rightarrow A_D$, where D is the number of decays and i is a dummy index ($i = $ 1, 2, 3, $...D$), each nuclide population can be found in terms of the previous population. In this case $N_2 = 0$, $N_3 = 0$,..., $N_D = 0$. Using the above result in a recursive form:

$$\frac{dN_j}{dt} = -\lambda_j N_j + \lambda_{j-1} N_{(j-1)0} e^{-\lambda_{j-1} t}.$$

The general solution to the recursive problem is given by *Bateman's equations*:[19]

Alternative decay modes

In all of the above examples, the initial nuclide decays into only one product.[20] Consider the case of one initial nuclide that can decay into either of two products, that is $A \rightarrow B$ and $A \rightarrow C$ in parallel. For example, in a sample of potassium-40, 89.3% of the nuclei decay to calcium-40 and 10.7% to argon-40. We have for all time t:

$$N = N_A + N_B + N_C$$

which is constant, since the total number of nuclides remains constant. Differentiating with respect to time:

$$\frac{dN_A}{dt} = -\left(\frac{dN_B}{dt} + \frac{dN_C}{dt}\right)$$
$$-\lambda N_A = -N_A\left(\lambda_B + \lambda_C\right)$$

defining the *total decay constant* λ in terms of the sum of *partial decay constants* λB and λC:

$$\lambda = \lambda_B + \lambda_C.$$

Notice that

$$\frac{dN_A}{dt} < 0, \frac{dN_B}{dt} > 0, \frac{dN_C}{dt} > 0.$$

Solving this equation for NA:

$$N_A = N_{A0}e^{-\lambda t}.$$

where NA_0 is the initial number of nuclide A. When measuring the production of one nuclide, one can only observe the total decay constant λ. The decay constants λB and λC determine the probability for the decay to result in products B or C as follows:

$$N_B = \frac{\lambda_B}{\lambda} N_{A0}\left(1 - e^{-\lambda t}\right),$$

$$N_C = \frac{\lambda_C}{\lambda} N_{A0}\left(1 - e^{-\lambda t}\right).$$

because the fraction $\lambda B/\lambda$ of nuclei decay into B while the fraction $\lambda C/\lambda$ of nuclei decay into C.

2.6.2 Corollaries of the decay laws

The above equations can also be written using quantities related to the number of nuclide particles N in a sample;

- The activity: $A = \lambda N$.

- The amount of substance: $n = N/L$.

- The mass: $M = Arn = ArN/L$.

where $L = 6.022 \times 10^{23}$ is Avogadro's constant, Ar is the relative atomic mass number, and the amount of the substance is in moles.

2.6.3 Decay timing: definitions and relations

Time constant and mean-life

For the one-decay solution $A \rightarrow B$:

$$N = N_0 \, e^{-\lambda t} = N_0 \, e^{-t/\tau},$$

the equation indicates that the decay constant λ has units of t^{-1}, and can thus also be represented as $1/\tau$, where τ is a characteristic time of the process called the *time constant*.

In a radioactive decay process, this time constant is also the mean lifetime for decaying atoms. Each atom "lives" for a finite amount of time before it decays, and it may be shown that this mean lifetime is the arithmetic mean of all the atoms' lifetimes, and that it is τ, which again is related to the decay constant as follows:

$$\tau = \frac{1}{\lambda}.$$

This form is also true for two-decay processes simultaneously $A \rightarrow B + C$, inserting the equivalent values of decay constants (as given above)

$$\lambda = \lambda_B + \lambda_C$$

into the decay solution leads to:

$$\frac{1}{\tau} = \lambda = \lambda_B + \lambda_C = \frac{1}{\tau_B} + \frac{1}{\tau_C}$$

Half-life

A more commonly used parameter is the half-life. Given a sample of a particular radionuclide, the half-life is the time taken for half the radionuclide's atoms to decay. For the case of one-decay nuclear reactions:

$$N = N_0 \, e^{-\lambda t} = N_0 \, e^{-t/\tau},$$

the half-life is related to the decay constant as follows: set $N = N_0/2$ and $t = T_{1/2}$ to obtain

$$t_{1/2} = \frac{\ln 2}{\lambda} = \tau \ln 2.$$

This relationship between the half-life and the decay constant shows that highly radioactive substances are quickly spent, while those that radiate weakly endure longer. Half-lives of known radionuclides vary widely, from more than 10^{19} years, such as for the very nearly stable nuclide ^{209}Bi, to 10^{-23} seconds for highly unstable ones.

The factor of $\ln(2)$ in the above relations results from the fact that concept of "half-life" is merely a way of selecting a different base other than the natural base e for the lifetime expression. The time constant τ is the $e - 1$ -life, the time until only $1/e$ remains, about 36.8%, rather than the 50% in the half-life of a radionuclide. Thus, τ is longer than $t_{1/2}$. The following equation can be shown to be valid:

$$N(t) = N_0 \, e^{-t/\tau} = N_0 \, 2^{-t/t_{1/2}}.$$

Since radioactive decay is exponential with a constant probability, each process could as easily be described with a different constant time period that (for example) gave its "(1/3)-life" (how long until only 1/3 is left) or "(1/10)-life" (a time period until only 10% is left), and so on. Thus, the choice of τ and *t1/2* for marker-times, are only for convenience, and from convention. They reflect a fundamental principle only in so much as they show that the *same proportion* of a given radioactive substance will decay, during any time-period that one chooses.

Mathematically, the nth life for the above situation would be found in the same way as above—by setting $N = N_0/n$, {{{1}}} and substituting into the decay solution to obtain

$$t_{1/n} = \frac{\ln n}{\lambda} = \tau \ln n.$$

2.6.4 Example

A sample of ^{14}C has a half-life of 5,730 years and a decay rate of 14 disintegration per minute (dpm) per gram of natural carbon.

If an artifact is found to have radioactivity of 4 dpm per gram of its present C, we can find the approximate age of the object using the above equation:

$$N = N_0 \, e^{-t/\tau},$$

where: $\frac{N}{N_0} = 4/14 \approx 0.286$,

$$\tau = \frac{T_{1/2}}{\ln 2} \approx 8267$$

$$t = -\tau \ln \frac{N}{N_0} \approx 10360$$

2.7 Changing decay rates

The radioactive decay modes of electron capture and internal conversion are known to be slightly sensitive to chemical and environmental effects that change the electronic structure of the atom, which in turn affects the presence of **1s** and **2s** electrons that participate in the decay process. A small number of mostly light nuclides are affected. For example, chemical bonds can affect the rate of electron capture to a small degree (in general, less than 1%) depending on the proximity of electrons to the nucleus. In ^{7}Be, a difference of 0.9% has been observed between half-lives in metallic and insulating environments.[21] This relatively large effect is because beryllium is a small atom whose valence electrons are in **2s** atomic orbitals, which are subject to electron capture in ^{7}Be because (like all **s** atomic orbitals in all atoms) they naturally penetrate into the nucleus.

In 1992, Jung et al. of the Darmstadt Heavy-Ion Research group observed an accelerated β decay of ^{163}Dy^{66+}. Although neutral ^{163}Dy is a stable isotope, the fully ionized ^{163}Dy^{66+} undergoes β decay into the K and L shells with a half-life of 47 days.[22]

Rhenium-187 is another spectacular example. ^{187}Re normally beta decays to ^{187}Os with a half-life of 41.6×10^{9} years,[23] but studies using fully ionised ^{187}Re atoms (bare nuclei) have found that this can decrease to only 33 years. This is attributed to "bound-state β^{-} decay" of the fully ionised atom – the electron is emitted into the "K-shell" (**1s** atomic orbital), which cannot occur for neutral atoms in which all low-lying bound states are occupied.[24]

A number of experiments have found that decay rates of other modes of artificial and naturally occurring radioisotopes are, to a high degree of precision, unaffected by external conditions such as temperature, pressure, the chemical environment, and electric, magnetic, or gravitational fields.[25] Comparison of laboratory experiments over the last century, studies of the Oklo natural nuclear reactor (which exemplified the effects of thermal neutrons on nuclear decay), and astrophysical observations of the luminosity decays of distant supernovae (which occurred far away so the light has taken a great deal of time to reach us), for example, strongly indicate that unperturbed decay rates have been constant (at least to within the limitations of small experimental errors) as a function of time as well.

Recent results suggest the possibility that decay rates might have a weak dependence on environmental factors. It has been suggested that measurements of decay rates of silicon-32, manganese-54, and radium-226 exhibit small seasonal variations (of the order of 0.1%),[26][27][28] while the decay of Radon-222 exhibit large 4% peak-to-peak seasonal variations,[29] proposed to be related to either solar flare activity or distance from the Sun. However, such measurements are highly susceptible to systematic errors, and a subsequent paper[30] has found no evidence for such correlations in seven other isotopes (^{22}Na, ^{44}Ti, ^{108}Ag, ^{121}Sn, ^{133}Ba, ^{241}Am, ^{238}Pu), and sets upper limits on the size of any such effects.

2.8   Theoretical basis of decay phenomena

The neutrons and protons that constitute nuclei, as well as other particles that approach close enough to them, are governed by several interactions. The strong nuclear force, not observed at the familiar macroscopic scale, is the most powerful force over subatomic distances. The electrostatic force is almost always significant, and, in the case of beta decay, the weak nuclear force is also involved.

The interplay of these forces produces a number of different phenomena in which energy may be released by rearrangement of particles in the nucleus, or else the change of one type of particle into others. These rearrangements and transformations may be hindered energetically, so that they do not occur immediately. In certain cases, random quantum vacuum fluctuations are theorized to promote relaxation to a lower energy state (the "decay") in a phenomenon known as quantum tunneling. Radioactive decay half-life of nuclides has been measured over timescales of 55 orders of magnitude, from 2.3 x 10^{-23} seconds (for hydrogen-7) to 6.9 x 10^{31} seconds (for tellurium-128).[31] The limits of these timescales are set by the sensitivity of instrumentation only, and there are no known natural limits to how brief or long a decay half life for radioactive decay of a radionuclide may be.

The decay process, like all hindered energy transformations, may be analogized by a snowfield on a mountain. While friction between the ice crystals may be supporting the snow's weight, the system is inherently unstable with regard to a state of lower potential energy. A disturbance would thus facilitate the path to a state of greater entropy: The system will move towards the ground state, producing heat, and the total energy will be distributable over a larger number of quantum states. Thus, an avalanche results. The *total* energy does not change in this process, but, because of the second law of thermodynamics, avalanches have only been observed in one direction and that is toward the "ground state" — the state with the largest number of ways in which the available energy could be distributed.

Such a collapse (a *decay event*) requires a specific activation energy. For a snow avalanche, this energy comes as a disturbance from outside the system, although such disturbances can be arbitrarily small. In the case of an excited atomic nucleus, the arbitrarily small disturbance comes from quantum vacuum fluctuations. A radioactive nucleus (or any excited system in quantum mechanics) is unstable, and can, thus, *spontaneously* stabilize to a less-excited system. The resulting transformation alters the structure of the nucleus and results in the emission of either a photon or a high-velocity particle that has mass (such as an electron, alpha particle, or other type).

2.9   Occurrence and applications

According to the Big Bang theory, stable isotopes of the lightest five elements (H, He, and traces of Li, Be, and B) were produced very shortly after the emergence of the universe, in a process called Big Bang nucleosynthesis. These lightest stable nuclides (including deuterium) survive to today, but any radioactive isotopes of the light elements produced in the Big Bang (such as tritium) have long since decayed. Isotopes of elements heavier than boron were not produced at all in the Big Bang, and these first five elements do not have any long-lived radioisotopes. Thus, all radioactive nuclei are, therefore, relatively young with respect to the birth of the universe, having formed later in various other types of nucleosynthesis in stars (in particular, supernovae), and also during ongoing interactions between stable isotopes and energetic particles. For example, carbon-14, a radioactive nuclide with a half-life of only 5,730 years, is constantly produced in Earth's upper atmosphere due to interactions between cosmic rays and nitrogen.

Nuclides that are produced by radioactive decay are called radiogenic nuclides, whether they themselves are stable or not. There exist stable radiogenic nuclides that were formed from short-lived extinct radionuclides in the early solar system.[32][33] The extra presence of these stable radiogenic nuclides (such as Xe-129 from primordial I-129) against the background of primordial stable nuclides can be inferred by various means.

Radioactive decay has been put to use in the technique of radioisotopic labeling, which is used to track the passage of a chemical substance through a complex system (such as a living organism). A sample of the substance is synthesized with a high concentration of unstable atoms. The presence of the substance in one or another part of the system is determined by detecting the locations of decay events.

On the premise that radioactive decay is truly random (rather than merely chaotic), it has been used in hardware random-number generators. Because the process is not thought to vary significantly in mechanism over time, it is also a valuable tool in estimating the absolute ages of certain materials. For geological materials, the radioisotopes and some of their decay products become trapped when a rock solidifies, and can then later be used (subject to many well-known qualifications) to estimate the date of the solidification. These include checking the results of several simultaneous processes and their products against each other, within the same sample. In a similar fashion, and also subject to qualification, the rate of formation of carbon-14 in various eras, the date of formation of organic matter within a certain period related to the isotope's half-life may be estimated, because the carbon-14 becomes trapped when the organic matter grows and incorporates the new carbon-14 from the air. Thereafter, the amount of carbon-14 in organic matter decreases according to decay processes that may also be independently cross-checked by other means (such as checking the carbon-14 in individual tree rings, for example).

2.10   Origins of radioactive nuclides

Main article: nucleosynthesis

Radioactive primordial nuclides found in the Earth are residues from ancient supernova explosions which occurred before the formation of the solar system. They are the long-lived fraction of radionuclides surviving in the primordial solar nebula through planet accretion until the present. The naturally occurring short-lived radiogenic radionuclides found in rocks are the daughters of these radioactive primordial nuclides. Another minor source of naturally occurring radioactive nuclides are cosmogenic nuclides, formed by cosmic ray bombardment of material in the Earth's atmosphere or crust. The radioactive decay of these radionuclides in rocks within Earth's mantle and crust contribute significantly to Earth's internal heat budget.

2.11   Decay chains and multiple modes

The daughter nuclide of a decay event may also be unstable (radioactive). In this case, it will also decay, producing radiation. The resulting second daughter nuclide may also be radioactive. This can lead to a sequence of several decay events. Eventually, a stable nuclide is produced. This is called a *decay chain* (see this article for specific details of important natural decay chains).

An example is the natural decay chain of ^{238}U, which is as follows:

- decays, through alpha-emission, with a half-life of 4.5 billion years to thorium-234

- which decays, through beta-emission, with a half-life of 24 days to protactinium-234

- which decays, through beta-emission, with a half-life of 1.2 minutes to uranium-234

- which decays, through alpha-emission, with a half-life of 240 thousand years to thorium-230

- which decays, through alpha-emission, with a half-life of 77 thousand years to radium-226

- which decays, through alpha-emission, with a half-life of 1.6 thousand years to radon-222

- which decays, through alpha-emission, with a half-life of 3.8 days to polonium-218

- which decays, through alpha-emission, with a half-life of 3.1 minutes to lead-214

- which decays, through beta-emission, with a half-life of 27 minutes to bismuth-214

- which decays, through beta-emission, with a half-life of 20 minutes to polonium-214

- which decays, through alpha-emission, with a half-life of 160 microseconds to lead-210

- which decays, through beta-emission, with a half-life of 22 years to bismuth-210

- which decays, through beta-emission, with a half-life of 5 days to polonium-210

- which decays, through alpha-emission, with a half-life of 140 days to lead-206, which is a stable nuclide.

Some radionuclides may have several different paths of decay. For example, approximately 36% of bismuth-212 decays, through alpha-emission, to thallium-208 while approximately 64% of bismuth-212 decays, through beta-emission, to polonium-212. Both thallium-208 and polonium-212 are radioactive daughter products of bismuth-212, and both decay directly to stable lead-208.

2.12   Associated hazard warning signs

- The trefoil symbol used to indicate ionising radiation.

- 2007 ISO radioactivity danger symbol intended for IAEA Category 1, 2 and 3 sources defined as dangerous sources capable of death or serious injury.[1]

- The dangerous goods transport classification sign for radioactive materials

1. ^ IAEA news release Feb 2007

2.13   See also

- Actinides in the environment

- Background radiation

- Chernobyl disaster

- Crimes involving radioactive substances

- Decay chain

- Fallout shelter

- Half-life

- Lists of nuclear disasters and radioactive incidents

- National Council on Radiation Protection and Measurements

- Nuclear engineering

- Nuclear medicine

- Nuclear pharmacy

- Nuclear physics

- Nuclear power

- Particle decay

- Poisson process

- Radiation

- Radiation therapy

- Radioactive contamination

- Radioactivity in biology

- Radiometric dating

- Radionuclide a.k.a. "radio-isotope"

- Secular equilibrium

- Transient equilibrium

2.14 Notes

[1] Radionuclide is the more correct term, but radioisotope is also used. The difference between isotope and nuclide is explained at Isotope#Isotope vs. nuclide.

2.15 References

2.15.1 Inline

[1] "Decay and Half Life". Retrieved 2009-12-14.

[2] Stabin, Michael G. (2007). "3". *Radiation Protection and Dosimetry: An Introduction to Health Physics.* Springer. doi:10.1007/978-0-387-49983-3. ISBN 978-0387499826.

[3] Best, Lara; Rodrigues, George; Velker, Vikram (2013). "1.3". *Radiation Oncology Primer and Review.* Demos Medical Publishing. ISBN 978-1620700044.

[4] Loveland, W.; Morrissey, D.; Seaborg, G.T. (2006). *Modern Nuclear Chemistry.* Wiley-Interscience. p. 57. ISBN 0-471-11532-0.

[5] Mould, Richard F. (1995). *A century of X-rays and radioactivity in medicine : with emphasis on photographic records of the early years* (Reprint. with minor corr ed.). Bristol: Inst. of Physics Publ. p. 12. ISBN 9780750302241.

[6] Kasimir Fajans, "Radioactive transformations and the periodic system of the elements". Berichte der Deutschen Chemischen Gesellschaft, Nr. 46, 1913, p. 422–439

[7] Frederick Soddy, "The Radio Elements and the Periodic Law", Chem. News, Nr. 107, 1913, p.97–99

[8] Sansare, K.; Khanna, V.; Karjodkar, F. (2011). "Early victims of X-rays: a tribute and current perception". *Dentomaxillofacial Radiology* **40** (2): 123–125. doi:10.1259/dmfr/73488299. ISSN 0250-832X. PMC 3520298. PMID 21239576.

[9] Ronald L. Kathern and Paul L. Ziemer, he First Fifty Years of Radiation Protection, physics.isu.edu

[10] Hrabak, M.; Padovan, R. S.; Kralik, M.; Ozretic, D.; Potocki, K. (July 2008). "Nikola Tesla and the Discovery of X-rays". *RadioGraphics* **28** (4): 1189–92. doi:10.1148/rg.284075206. PMID 18635636.

[11] Geoff Meggitt (2008), *Taming the Rays - A history of Radiation and Protection.*, Lulu.com, ISBN 978-1-4092-4667-1

[12] Clarke, R.H.; J. Valentin (2009). "The History of ICRP and the Evolution of its Policies" (PDF). *Annals of the ICRP*. ICRP Publication 109 **39** (1): pp. 75–110. doi:10.1016/j.icrp.2009.07.009. Retrieved 12 May 2012.

[13] Rutherford, Ernest (6 October 1910). "Radium Standards and Nomenclature". *Nature* **84** (2136): 430–431.

[14] *10 CFR 20.1005*. US Nuclear Regulatory Commission. 2009.

[15] The Council of the European Communities (1979-12-21). "Council Directive 80/181/EEC of 20 December 1979 on the approximation of the laws of the Member States relating to Unit of measurement and on the repeal of Directive 71/354/EEC". Retrieved 19 May 2012.

[16] Radioactive Decay

[17] Patel, S.B. (2000). *Nuclear physics : an introduction*. New Delhi: New Age International. pp. 62–72. ISBN 9788122401257.

[18] Introductory Nuclear Physics, K.S. Krane, 1988, John Wiley & Sons Inc, ISBN 978-0-471-80553-3

[19] Cetnar, Jerzy (May 2006). "General solution of Bateman equations for nuclear transmutations". *Annals of Nuclear Energy* **33** (7): 640–645. doi:10.1016/j.anucene.2006.02.004.

[20] K.S. Krane (1988). *Introductory Nuclear Physics*. John Wiley & Sons Inc. p. 164. ISBN 978-0-471-80553-3.

[21] Wang, B.; Yan, S.; Limata, B. et al. (2006). "Change of the 7Be electron capture half-life in metallic environments". *The European Physical Journal A* **28** (3): 375–377. Bibcode:2006EPJA...28..375W. doi:10.1140/epja/i2006-10068-x. ISSN 1434-6001.

[22] Jung, M.; Bosch, F.; Beckert, K. et al. (1992). "First observation of bound-state β⁻ decay". *Physical Review Letters* **69** (15): 2164–2167. Bibcode:1992PhRvL..69.2164J. doi:10.1103/PhysRevLett.69.2164. ISSN 0031-9007. PMID 10046415.

[23] Smoliar, M.I.; Walker, R.J.; Morgan, J.W. (1996). "Re-Os ages of group IIA, IIIA, IVA, and IVB iron meteorites". *Science* **271** (5252): 1099–1102. Bibcode:1996Sci...271.1099S. doi:10.1126/science.271.5252.1099.

[24] Bosch, F.; Faestermann, T.; Friese, J.; Heine, F.; Kienle, P.; Wefers, E.; Zeitelhack, K.; Beckert, K.; Franzke, B.; Klepper, O.; Kozhuharov, C.; Menzel, G.; Moshammer, R.; Nolden, F.; Reich, H.; Schlitt, B.; Steck, M.; Stöhlker, T.; Winkler, T.; Takahashi, K. (1996). "Observation of bound-state β– decay of fully ionized ^{187}Re:^{187}Re-^{187}Os Cosmochronometry". *Physical Review Letters* **77** (26): 5190–5193. Bibcode:1996PhRvL..77.5190B. doi:10.1103/PhysRevLett.77.5190. PMID 10062738.

[25] Emery, G.T. (1972). "Perturbation of Nuclear Decay Rates" (PDF). *Annual Review of Nuclear Science* (ACS Publications) **22**: 165–202. Bibcode:1972ARNPS..22..165E. doi:10.1146/annurev.ns.22.120172.001121. Retrieved 6 August 2012.

[26] "The mystery of varying nuclear decay". *Physics World*. 2 October 2008.

[27] Jenkins, Jere H.; Fischbach, Ephraim (2009). "Perturbation of Nuclear Decay Rates During the Solar Flare of 13 December 2006".*Astroparticle Physics*31(6):407–411.arXiv:0808.3156.Bibcode:2009APh....31..407J.doi:10.1016/j.astropartphys.2005.

[28] Jenkins, J. H.; Buncher, John B.; Gruenwald, John T.; Krause, Dennis E.; Mattes, Joshua J. et al. (2009). "Evidence of correlations between nuclear decay rates and Earth–Sun distance". *Astroparticle Physics* **32** (1): 42–46. arXiv:0808.3283. Bibcode:2009APh....32...42J. doi:10.1016/j.astropartphys.2009.05.004.

[29] Peter A. Sturrock, Gideon Steinitz, Ephraim Fischbach, Daniel Javorsek, II, Jere H. Jenkins, Analysis of Gamma Radiation from a Radon Source: Indications of a Solar Influence, Accessed on line September 2, 2012.

[30] Norman, E. B.; Shugart, Howard A.; Joshi, Tenzing H.; Firestone, Richard B. et al. (2009). "Evidence against correlations between nuclear decay rates and Earth–Sun distance" (PDF). *Astroparticle Physics* **31** (2): 135–137. arXiv:0810.3265. Bibcode:2009APh....31..135N. doi:10.1016/j.astropartphys.2008.12.004.

[31] NUBASE evaluation of nuclear and decay properties

[32] Clayton, Donald D. (1983). *Principles of Stellar Evolution and Nucleosynthesis* (2nd ed.). University of Chicago Press. p. 75. ISBN 0-226-10953-4.

[33] Bolt, B. A.; Packard, R. E.; Price, P. B. (2007). "John H. Reynolds, Physics: Berkeley". The University of California, Berkeley. Retrieved 2007-10-01.

2.15.2 General

- "Radioactivity", Encyclopædia Britannica. 2006. Encyclopædia Britannica Online. December 18, 2006

- Radio-activity by Ernest Rutherford Phd, Encyclopædia Britannica Eleventh Edition

2.16   External links

- The Lund/LBNL Nuclear Data Search – Contains tabulated information on radioactive decay types and energies.

- Nomenclature of nuclear chemistry

- Specific activity and related topics.

- The Live Chart of Nuclides – IAEA

- Health Physics Society Public Education Website

- Beach, Chandler B., ed. (1914). "Becquerel Rays". *The New Student's Reference Work*. Chicago: F. E. Compton and Co.

- Annotated bibliography for radioactivity from the Alsos Digital Library for Nuclear Issues

- Stochastic Java applet on the decay of radioactive atoms by Wolfgang Bauer

- Stochastic Flash simulation on the decay of radioactive atoms by David M. Harrison

- "Henri Becquerel: The Discovery of Radioactivity", Becquerel's 1896 articles online and analyzed on *BibNum* [click 'à télécharger' for English version].

- "Radioactive change", Rutherford & Soddy article (1903), online and analyzed on *Bibnum* [click 'à télécharger' for English version].

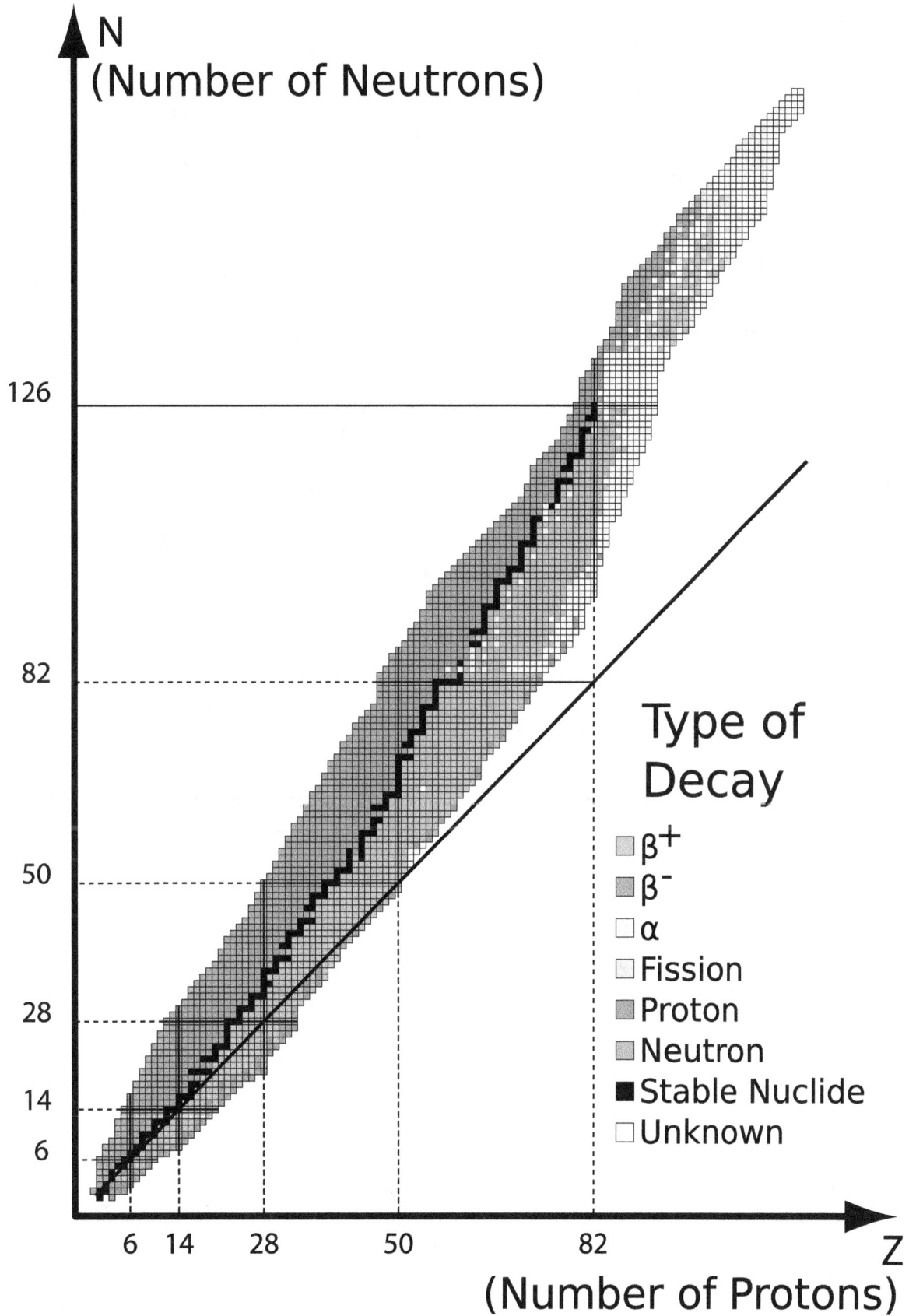

Types of radioactive decay related to N and Z numbers

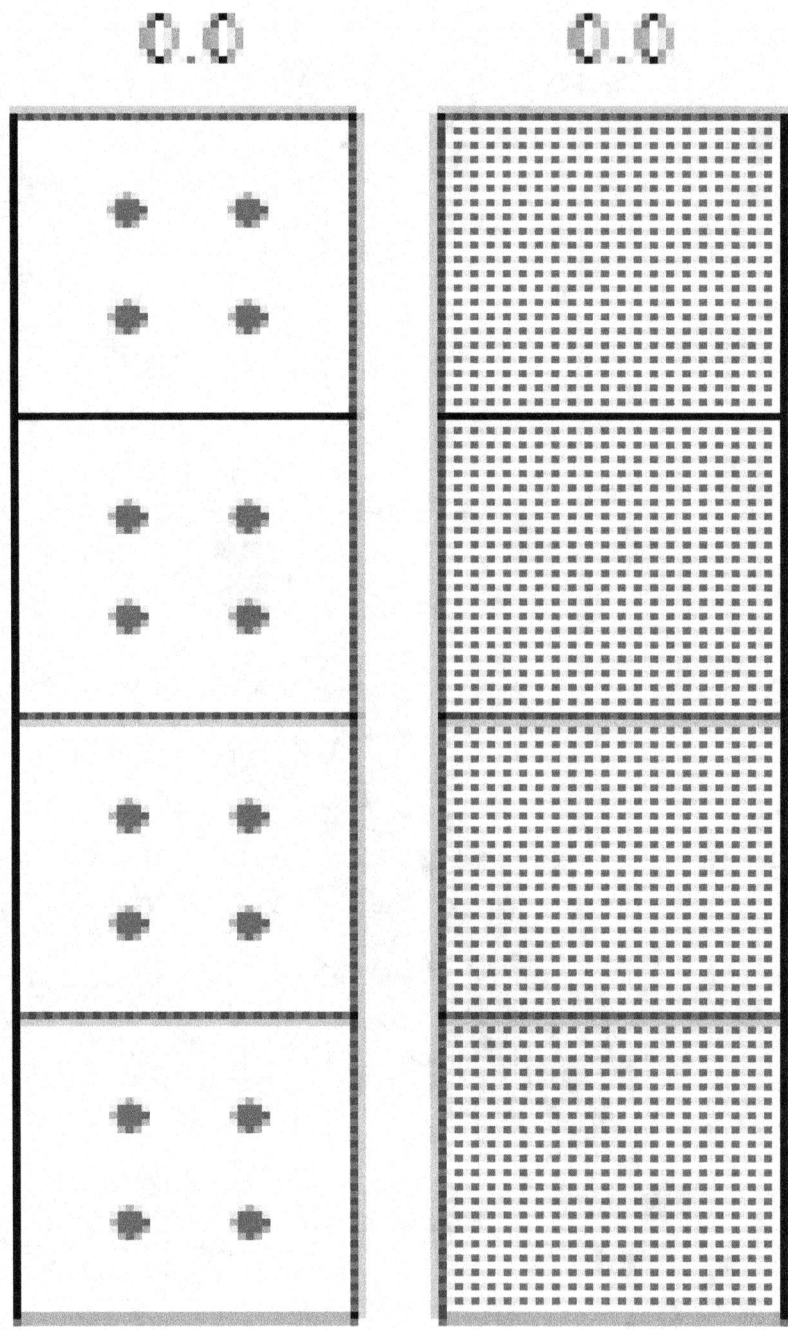

Simulation of many identical atoms undergoing radioactive decay, starting with either 4 atoms (left) or 400 (right). The number at the top indicates how many half-lives have elapsed. Note the law of large numbers: with more atoms, the overall decay is less random
.

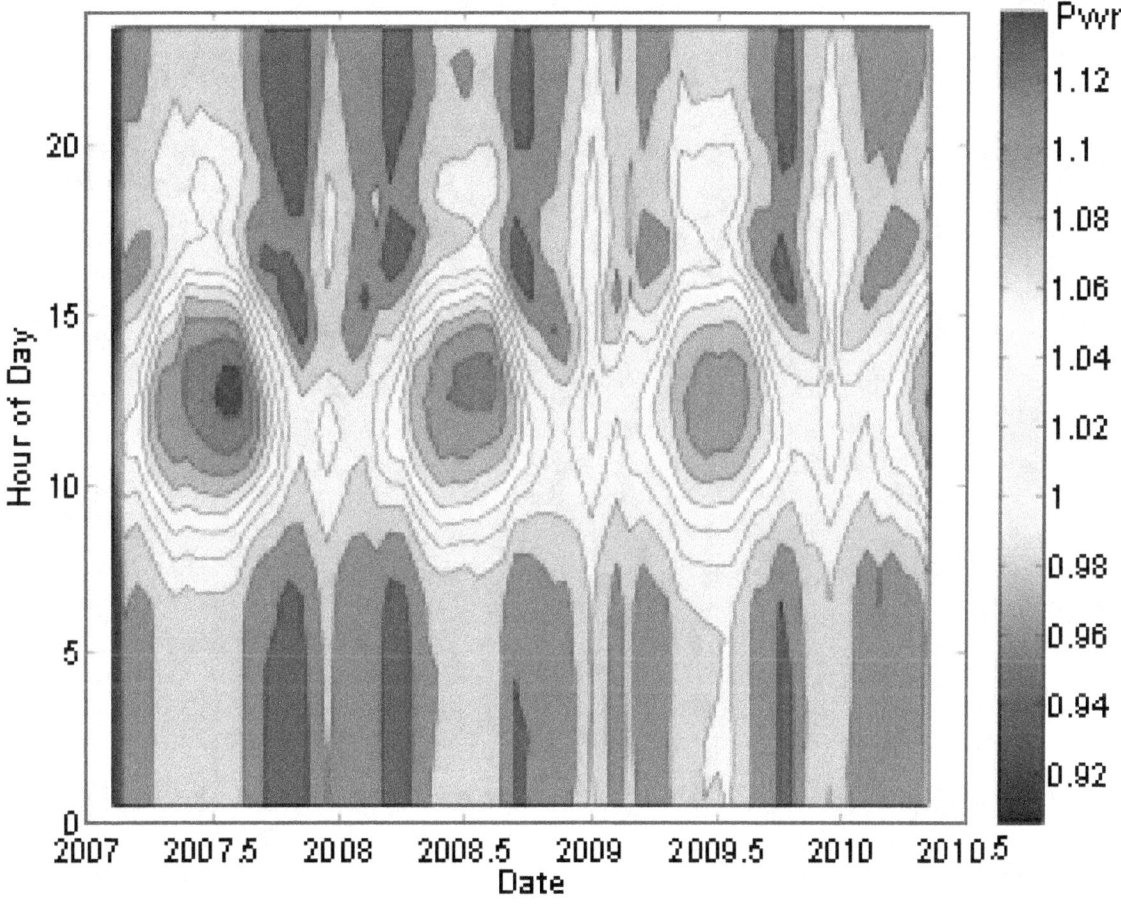

Decay Rate of Radon-222 as a function of date and time of day. The color-bar gives the power of the observed signal and represents ~4% seasonal decay rate variation.

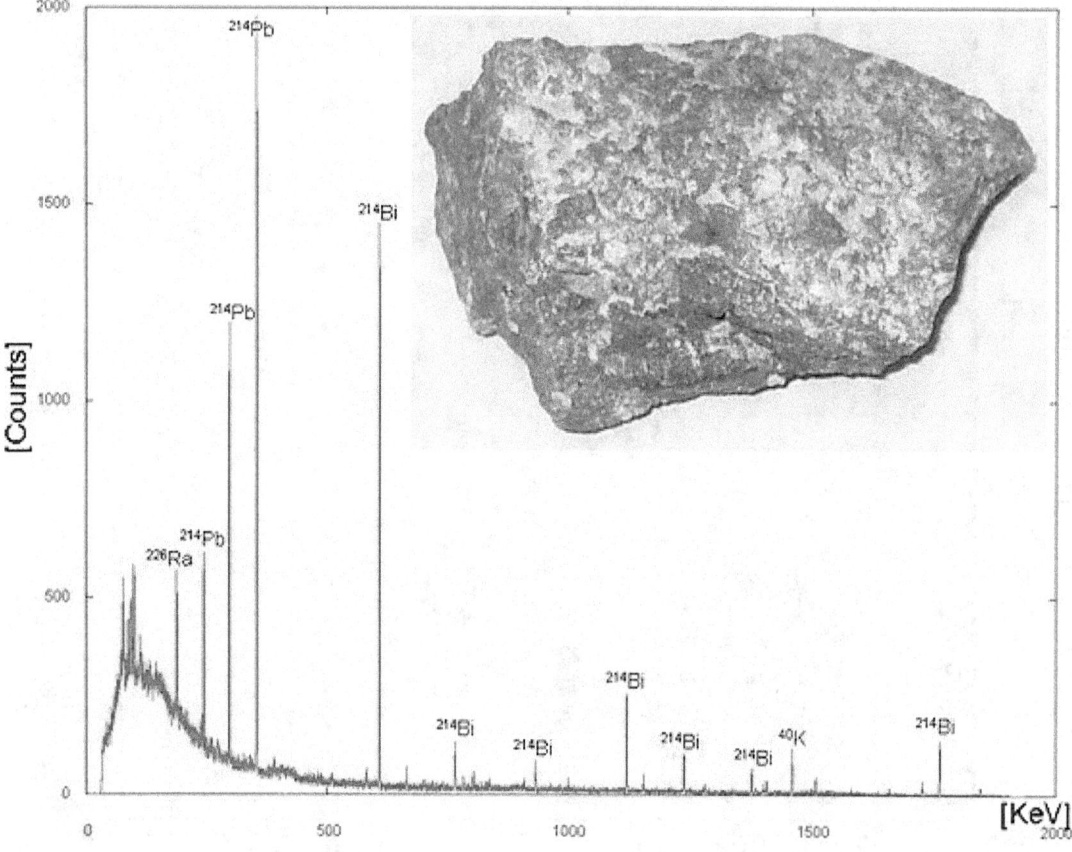

Gamma-ray energy spectrum of uranium ore (inset). Gamma-rays are emitted by decaying nuclides, and the gamma-ray energy can be used to characterize the decay (which nuclide is decaying to which). Here, using the gamma-ray spectrum, several nuclides that are typical of the decay chain of ^{238}U have been identified: ^{226}Ra, ^{214}Pb, ^{214}Bi.

Chapter 3

Nuclear fission

"Splitting the atom" redirects here. For the EP, see Splitting the Atom.

In nuclear physics and nuclear chemistry, **nuclear fission** is either a nuclear reaction or a radioactive decay process in which the nucleus of an atom splits into smaller parts (lighter nuclei). The fission process often produces free neutrons and photons (in the form of gamma rays), and releases a very large amount of energy even by the energetic standards of radioactive decay.

Nuclear fission of heavy elements was discovered on December 17, 1938 by German Otto Hahn and his assistant Fritz Strassmann, and explained theoretically in January 1939 by Lise Meitner and her nephew Otto Robert Frisch. Frisch named the process by analogy with biological fission of living cells. It is an exothermic reaction which can release large amounts of energy both as electromagnetic radiation and as kinetic energy of the fragments (heating the bulk material where fission takes place). In order for fission to produce energy, the total binding energy of the resulting elements must be less negative (higher energy) than that of the starting element.

Fission is a form of nuclear transmutation because the resulting fragments are not the same element as the original atom. The two nuclei produced are most often of comparable but slightly different sizes, typically with a mass ratio of products of about 3 to 2, for common fissile isotopes.[1][2] Most fissions are binary fissions (producing two charged fragments), but occasionally (2 to 4 times per 1000 events), *three* positively charged fragments are produced, in a ternary fission. The smallest of these fragments in ternary processes ranges in size from a proton to an argon nucleus.

Apart from fission induced by a neutron, harnessed and exploited by humans, a natural form of spontaneous radioactive decay (not requiring a neutron) is also referred to as fission, and occurs especially in very high-mass-number isotopes. Spontaneous fission was discovered in 1940 by Flyorov, Petrzhak and Kurchatov[3] in Moscow, when they decided to confirm that, without bombardment by neutrons, the fission rate of uranium was indeed negligible, as predicted by Niels Bohr; it wasn't.[3]

The unpredictable composition of the products (which vary in a broad probabilistic and somewhat chaotic manner) distinguishes fission from purely quantum-tunnelling processes such as proton emission, alpha decay and cluster decay, which give the same products each time. Nuclear fission produces energy for nuclear power and drives the explosion of nuclear weapons. Both uses are possible because certain substances called nuclear fuels undergo fission when struck by fission neutrons, and in turn emit neutrons when they break apart. This makes possible a self-sustaining nuclear chain reaction that releases energy at a controlled rate in a nuclear reactor or at a very rapid uncontrolled rate in a nuclear weapon.

The amount of free energy contained in nuclear fuel is millions of times the amount of free energy contained in a similar mass of chemical fuel such as gasoline, making nuclear fission a very dense source of energy. The products of nuclear fission, however, are on average far more radioactive than the heavy elements which are normally fissioned as fuel, and remain so for significant amounts of time, giving rise to a nuclear waste problem. Concerns over nuclear waste accumulation and over the destructive potential of nuclear weapons may counterbalance the desirable qualities of fission as an energy source, and give rise to ongoing political debate over nuclear power.

3.1 Physical overview

3.1.1 Mechanism

Nuclear fission can occur without neutron bombardment as a type of radioactive decay. This type of fission (called spontaneous fission) is rare except in a few heavy isotopes. In engineered nuclear devices, essentially all nuclear fission occurs as a "nuclear reaction" — a bombardment-driven process that results from the collision of two subatomic particles. In nuclear reactions, a subatomic particle collides with an atomic nucleus and causes changes to it. Nuclear reactions are thus driven by the mechanics of bombardment, not by the relatively constant exponential decay and half-life characteristic of spontaneous radioactive processes.

Many types of nuclear reactions are currently known. Nuclear fission differs importantly from other types of nuclear reactions, in that it can be amplified and sometimes controlled via a nuclear chain reaction (one type of general chain reaction). In such a reaction, free neutrons released by each fission event can trigger yet more events, which in turn release more neutrons and cause more fissions.

The chemical element isotopes that can sustain a fission chain reaction are called nuclear fuels, and are said to be *fissile*. The most common nuclear fuels are ^{235}U (the isotope of uranium with an atomic mass of 235 and of use in nuclear reactors) and ^{239}Pu (the isotope of plutonium with an atomic mass of 239). These fuels break apart into a bimodal range of chemical elements with atomic masses centering near 95 and 135 **u** (fission products). Most nuclear fuels undergo spontaneous fission only very slowly, decaying instead mainly via an alpha/beta decay chain over periods of millennia to eons. In a nuclear reactor or nuclear weapon, the overwhelming majority of fission events are induced by bombardment with another particle, a neutron, which is itself produced by prior fission events.

Nuclear fissions in fissile fuels are the result of the nuclear excitation energy produced when a fissile nucleus captures a neutron. This energy, resulting from the neutron capture, is a result of the attractive nuclear force acting between the neutron and nucleus. It is enough to deform the nucleus into a double-lobed "drop," to the point that nuclear fragments exceed the distances at which the nuclear force can hold two groups of charged nucleons together, and when this happens, the two fragments complete their separation and then are driven further apart by their mutually repulsive charges, in a process which becomes irreversible with greater and greater distance. A similar process occurs in fissionable isotopes (such as uranium-238), but in order to fission, these isotopes require additional energy provided by fast neutrons (such as those produced by nuclear fusion in thermonuclear weapons).

The liquid drop model of the atomic nucleus predicts equal-sized fission products as an outcome of nuclear deformation. The more sophisticated nuclear shell model is needed to mechanistically explain the route to the more energetically favorable outcome, in which one fission product is slightly smaller than the other. A theory of the fission based on shell model has been formulated by Maria Goeppert Mayer.

The most common fission process is binary fission, and it produces the fission products noted above, at 95±15 and 135±15 **u**. However, the binary process happens merely because it is the most probable. In anywhere from 2 to 4 fissions per 1000 in a nuclear reactor, a process called ternary fission produces three positively charged fragments (plus neutrons) and the smallest of these may range from so small a charge and mass as a proton (Z=1), to as large a fragment as argon (Z=18). The most common small fragments, however, are composed of 90% helium-4 nuclei with more energy than alpha particles from alpha decay (so-called "long range alphas" at ~ 16 MeV), plus helium-6 nuclei, and tritons (the nuclei of tritium). The ternary process is less common, but still ends up producing significant helium-4 and tritium gas buildup in the fuel rods of modern nuclear reactors.[4]

3.1.2 Energetics

Input

The fission of a heavy nucleus requires a total input energy of about 7 to 8 million electron volts (MeV) to initially overcome the nuclear force which holds the nucleus into a spherical or nearly spherical shape, and from there, deform it into a two-lobed ("peanut") shape in which the lobes are able to continue to separate from each other, pushed by their mutual positive charge, in the most common process of binary fission (two positively charged fission products + neutrons). Once the nuclear lobes have been pushed to a critical distance, beyond which the short range strong force can no longer hold them together, the process of their separation proceeds from the energy of the (longer range) electromagnetic repulsion between the fragments. The result is two fission fragments moving away from each other, at high energy.

About 6 MeV of the fission-input energy is supplied by the simple binding of an extra neutron to the heavy nucleus via the strong force; however, in many fissionable isotopes, this amount of energy is not enough for fission. Uranium-238, for example, has a near-zero fission cross section for neutrons of less than one MeV energy. If no additional energy is supplied by any other mechanism, the nucleus will not fission, but will merely absorb the neutron, as happens when U-238 absorbs slow and even some fraction of fast neutrons, to become U-239. The remaining energy to initiate fission can be supplied by two other mechanisms: one of these is more kinetic energy of the incoming neutron, which is increasingly able to fission a fissionable heavy nucleus as it exceeds a kinetic energy of one MeV or more (so-called fast neutrons). Such high energy neutrons are able to fission U-238 directly (see thermonuclear weapon for application, where the fast neutrons are supplied by nuclear fusion). However, this process cannot happen to a great extent in a nuclear reactor, as too small a fraction of the fission neutrons produced by any type of fission have enough energy to efficiently fission U-238 (fission neutrons have a mode energy of 2 MeV, but a median of only 0.75 MeV, meaning half of them have less than this insufficient energy).[5]

Among the heavy actinide elements, however, those isotopes that have an odd number of neutrons (such as U-235 with 143 neutrons) bind an extra neutron with an additional 1 to 2 MeV of energy over an isotope of the same element with an even number of neutrons (such as U-238 with 146 neutrons). This extra binding energy is made available as a result of the mechanism of neutron pairing effects. This extra energy results from the Pauli exclusion principle allowing an extra neutron to occupy the same nuclear orbital as the last neutron in the nucleus, so that the two form a pair. In such isotopes, therefore, no neutron kinetic energy is needed, for all the necessary energy is supplied by absorption of any neutron, either of the slow or fast variety (the former are used in moderated nuclear reactors, and the latter are used in fast neutron reactors, and in weapons). As noted above, the subgroup of fissionable elements that may be fissioned efficiently with their own fission neutrons (thus potentially causing a nuclear chain reaction in relatively small amounts of the pure material) are termed "fissile." Examples of fissile isotopes are U-235 and plutonium-239.

Output

Typical fission events release about two hundred million eV (200 MeV) of energy for each fission event. The exact isotope which is fissioned, and whether or not it is fissionable or fissile, has only a small impact on the amount of energy released. This can be easily seen by examining the curve of binding energy (image below), and noting that the average binding energy of the actinide nuclides beginning with uranium is around 7.6 MeV per nucleon. Looking further left on the curve of binding energy, where the fission products cluster, it is easily observed that the binding energy of the fission products tends to center around 8.5 MeV per nucleon. Thus, in any fission event of an isotope in the actinide's range of mass, roughly 0.9 MeV is released per nucleon of the starting element. The fission of U235 by a slow neutron yields nearly identical energy to the fission of U238 by a fast neutron. This energy release profile holds true for thorium and the various minor actinides as well.[6]

By contrast, most chemical oxidation reactions (such as burning coal or TNT) release at most a few eV per event. So, nuclear fuel contains at least ten million times more usable energy per unit mass than does chemical fuel. The energy of nuclear fission is released as kinetic energy of the fission products and fragments, and as electromagnetic radiation in the form of gamma rays; in a nuclear reactor, the energy is converted to heat as the particles and gamma rays collide with the atoms that make up the reactor and its working fluid, usually water or occasionally heavy water or molten salts.

When a uranium nucleus fissions into two daughter nuclei fragments, about 0.1 percent of the mass of the uranium nucleus[7] appears as the fission energy of ~200 MeV. For uranium-235 (total mean fission energy 202.5 MeV), typically ~169 MeV appears as the kinetic energy of the daughter nuclei, which fly apart at about 3% of the speed of light, due to Coulomb repulsion. Also, an average of 2.5 neutrons are emitted, with a mean kinetic energy per neutron of ~2 MeV (total of 4.8 MeV).[8] The fission reaction also releases ~7 MeV in prompt gamma ray photons. The latter figure means that a nuclear fission explosion or criticality accident emits about 3.5% of its energy as gamma rays, less than 2.5% of its energy as fast neutrons (total of both types of radiation ~ 6%), and the rest as kinetic energy of fission fragments (this appears almost immediately when the fragments impact surrounding matter, as simple heat). In an atomic bomb, this heat may serve to raise the temperature of the bomb core to 100 million kelvin and cause secondary emission of soft X-rays, which convert some of this energy to ionizing radiation. However, in nuclear reactors, the fission fragment kinetic energy remains as low-temperature heat, which itself causes little or no ionization.

So-called neutron bombs (enhanced radiation weapons) have been constructed which release a larger fraction of their energy as ionizing radiation (specifically, neutrons), but these are all thermonuclear devices which rely on the nuclear fusion stage to produce the extra radiation. The energy dynamics of pure fission bombs always remain at about 6% yield of the total in radiation, as a prompt result of fission.

The total *prompt fission* energy amounts to about 181 MeV, or ~ 89% of the total energy which is eventually released by fission over time. The remaining ~ 11% is released in beta decays which have various half-lives, but begin as a process in the fission products immediately; and in delayed gamma emissions associated with these beta decays. For example, in uranium-235 this delayed energy is divided into about 6.5 MeV in betas, 8.8 MeV in antineutrinos (released at the same time as the betas), and finally, an additional 6.3 MeV in delayed gamma emission from the excited beta-decay products (for a mean total of ~10 gamma ray emissions per fission, in all). Thus, about 6.5% of the total energy of fission is released some time after the event, as non-prompt or delayed ionizing radiation, and the delayed ionizing energy is about evenly divided between gamma and beta ray energy.

In a reactor that has been operating for some time, the radioactive fission products will have built up to steady state concentrations such that their rate of decay is equal to their rate of formation, so that their fractional total contribution to reactor heat (via beta decay) is the same as these radioisotopic fractional contributions to the energy of fission. Under these conditions, the 6.5% of fission which appears as delayed ionizing radiation (delayed gammas and betas from radioactive fission products) contributes to the steady-state reactor heat production under power. It is this output fraction which remains when the reactor is suddenly shut down (undergoes scram). For this reason, the reactor decay heat output begins at 6.5% of the full reactor steady state fission power, once the reactor is shut down. However, within hours, due to decay of these isotopes, the decay power output is far less. See decay heat for detail.

The remainder of the delayed energy (8.8 MeV/202.5 MeV = 4.3% of total fission energy) is emitted as antineutrinos, which as a practical matter, are not considered "ionizing radiation." The reason is that energy released as antineutrinos is not captured by the reactor material as heat, and escapes directly through all materials (including the Earth) at nearly the speed of light, and into interplanetary space (the amount absorbed is minuscule). Neutrino radiation is ordinarily not classed as ionizing radiation, because it is almost entirely not absorbed and therefore does not produce effects (although the very rare neutrino event is ionizing). Almost all of the rest of the radiation (6.5% delayed beta and gamma radiation) is eventually converted to heat in a reactor core or its shielding.

Some processes involving neutrons are notable for absorbing or finally yielding energy — for example neutron kinetic energy does not yield heat immediately if the neutron is captured by a uranium-238 atom to breed plutonium-239, but this energy is emitted if the plutonium-239 is later fissioned. On the other hand, so-called delayed neutrons emitted as radioactive decay products with half-lives up to several minutes, from fission-daughters, are very important to reactor control, because they give a characteristic "reaction" time for the total nuclear reaction to double in size, if the reaction is run in a "delayed-critical" zone which deliberately relies on these neutrons for a supercritical chain-reaction (one in which each fission cycle yields more neutrons than it absorbs). Without their existence, the nuclear chain-reaction would be prompt critical and increase in size faster than it could be controlled by human intervention. In this case, the first experimental atomic reactors would have run away to a dangerous and messy "prompt critical reaction" before their operators could have manually shut them down (for this reason, designer Enrico Fermi included radiation-counter-triggered control rods, suspended by electromagnets, which could automatically drop into the center of Chicago Pile-1). If these delayed neutrons are captured without producing fissions, they produce heat as well.[9]

3.1.3   Product nuclei and binding energy

Main articles: fission product and fission product yield

In fission there is a preference to yield fragments with even proton numbers, which is called the odd-even effect on the fragments charge distribution. However, no odd-even effect is observed on fragment **mass number** distribution. This result is attributed to nucleon pair breaking.

In nuclear fission events the nuclei may break into any combination of lighter nuclei, but the most common event is not fission to equal mass nuclei of about mass 120; the most common event (depending on isotope and process) is a slightly unequal fission in which one daughter nucleus has a mass of about 90 to 100 **u** and the other the remaining 130 to 140 **u**.[10] Unequal fissions are energetically more favorable because this allows one product to be closer to the energetic minimum near mass 60 **u** (only a quarter of the average fissionable mass), while the other nucleus with mass 135 **u** is still not far out of the range of the most tightly bound nuclei (another statement of this, is that the atomic binding energy curve is slightly steeper to the left of mass 120 **u** than to the right of it).

3.1.4   Origin of the active energy and the curve of binding energy

Nuclear fission of heavy elements produces energy because the specific binding energy (binding energy per mass) of intermediate-mass nuclei with atomic numbers and atomic masses close to ^{62}Ni and ^{56}Fe is greater than the nucleon-specific binding energy of very heavy nuclei, so that energy is released when heavy nuclei are broken apart. The total rest masses of the fission products (**Mp**) from a single reaction is less than the mass of the original fuel nucleus (**M**). The excess mass $\Delta m = M - Mp$ is the invariant mass of the energy that is released as photons (gamma rays) and kinetic energy of the fission fragments, according to the mass-energy equivalence formula $E = mc^2$.

The variation in specific binding energy with atomic number is due to the interplay of the two fundamental forces acting on the component nucleons (protons and neutrons) that make up the nucleus. Nuclei are bound by an attractive nuclear force between nucleons, which overcomes the electrostatic repulsion between protons. However, the nuclear force acts only over relatively short ranges (a few nucleon diameters), since it follows an exponentially decaying Yukawa potential which makes it insignificant at longer distances. The electrostatic repulsion is of longer range, since it decays by an inverse-square rule, so that nuclei larger than about 12 nucleons in diameter reach a point that the total electrostatic repulsion overcomes the nuclear force and causes them to be spontaneously unstable. For the same reason, larger nuclei (more than about eight nucleons in diameter) are less tightly bound per unit mass than are smaller nuclei; breaking a large nucleus into two or more intermediate-sized nuclei releases energy. The origin of this energy is the nuclear force, which intermediate-sized nuclei allows to act more efficiently, because each nucleon has more neighbors which are within the short range attraction of this force. Thus less energy is needed in the smaller nuclei and the difference to the state before is set free.

Also because of the short range of the strong binding force, large stable nuclei must contain proportionally more neutrons than do the lightest elements, which are most stable with a **1 to 1 ratio** of protons and neutrons. Nuclei which have more than 20 protons cannot be stable unless they have more than an equal number of neutrons. Extra neutrons stabilize heavy elements because they add to strong-force binding (which acts between all nucleons) without adding to proton–proton repulsion. Fission products have, on average, about the same ratio of neutrons and protons as their parent nucleus, and are therefore usually unstable to beta decay (which changes neutrons to protons) because they have proportionally too many neutrons compared to stable isotopes of similar mass.

This tendency for fission product nuclei to beta-decay is the fundamental cause of the problem of radioactive high level waste from nuclear reactors. Fission products tend to be beta emitters, emitting fast-moving electrons to conserve electric charge, as excess neutrons convert to protons in the fission-product atoms. See Fission products (by element) for a description of fission products sorted by element.

3.1.5   Chain reactions

Main article: Nuclear chain reaction

Several heavy elements, such as uranium, thorium, and plutonium, undergo both spontaneous fission, a form of radioactive decay and *induced fission*, a form of nuclear reaction. Elemental isotopes that undergo induced fission when struck by a free neutron are called fissionable; isotopes that undergo fission when struck by a thermal, slow moving neutron are also called fissile. A few particularly fissile and readily obtainable isotopes (notably ^{233}U, ^{235}U and ^{239}Pu) are called nuclear fuels because they can sustain a chain reaction and can be obtained in large enough quantities to be useful.

All fissionable and fissile isotopes undergo a small amount of spontaneous fission which releases a few free neutrons into any sample of nuclear fuel. Such neutrons would escape rapidly from the fuel and become a free neutron, with a mean lifetime of about 15 minutes before decaying to protons and beta particles. However, neutrons almost invariably impact and are absorbed by other nuclei in the vicinity long before this happens (newly created fission neutrons move at about 7% of the speed of light, and even moderated neutrons move at about 8 times the speed of sound). Some neutrons will impact fuel nuclei and induce further fissions, releasing yet more neutrons. If enough nuclear fuel is assembled in one place, or if the escaping neutrons are sufficiently contained, then these freshly emitted neutrons outnumber the neutrons that escape from the assembly, and a *sustained nuclear chain reaction* will take place.

An assembly that supports a sustained nuclear chain reaction is called a critical assembly or, if the assembly is almost entirely made of a nuclear fuel, a critical mass. The word "critical" refers to a cusp in the behavior of the differential equation that governs the number of free neutrons present in the fuel: if less than a critical mass is present, then the amount of neutrons is determined by radioactive decay, but if a critical mass or more is present, then the amount of neutrons is controlled instead by the physics of the chain reaction. The actual mass of a *critical mass* of nuclear fuel depends strongly on the geometry and surrounding materials.

Not all fissionable isotopes can sustain a chain reaction. For example, ^{238}U, the most abundant form of uranium, is fissionable but not fissile: it undergoes induced fission when impacted by an energetic neutron with over 1 MeV of kinetic energy. However, too few of the neutrons produced by ^{238}U fission are energetic enough to induce further fissions in ^{238}U, so no chain reaction is possible with this isotope. Instead, bombarding ^{238}U with slow neutrons causes it to absorb them (becoming ^{239}U) and decay by beta emission to ^{239}Np which then decays again by the same process to ^{239}Pu; that process is used to manufacture ^{239}Pu in breeder reactors. In-situ plutonium production also contributes to the neutron chain reaction in other types of reactors after sufficient plutonium-239 has been produced, since plutonium-239 is also a fissile element which serves as fuel. It is estimated that up to half of the power produced by a standard "non-breeder" reactor is produced by the fission of plutonium-239 produced in place, over the total life-cycle of a fuel load.

Fissionable, non-fissile isotopes can be used as fission energy source even without a chain reaction. Bombarding ^{238}U with fast neutrons induces fissions, releasing energy as long as the external neutron source is present. This is an important effect in all reactors where fast neutrons from the fissile isotope can cause the fission of nearby ^{238}U nuclei, which means that some small part of the ^{238}U is "burned-up" in all nuclear fuels, especially in fast breeder reactors that operate with higher-energy neutrons. That same fast-fission effect is used to augment the energy released by modern thermonuclear weapons, by jacketing the weapon with ^{238}U to react with neutrons released by nuclear fusion at the center of the device. But the explosive effects of nuclear fission chain reactions can be reduced by using substances like moderators which slow down the speed of secondary neutrons.[11]

3.1.6 Fission reactors

Critical fission reactors are the most common type of nuclear reactor. In a critical fission reactor, neutrons produced by fission of fuel atoms are used to induce yet more fissions, to sustain a controllable amount of energy release. Devices that produce engineered but non-self-sustaining fission reactions are subcritical fission reactors. Such devices use radioactive decay or particle accelerators to trigger fissions.

Critical fission reactors are built for three primary purposes, which typically involve different engineering trade-offs to take advantage of either the heat or the neutrons produced by the fission chain reaction:

- *power reactors* are intended to produce heat for nuclear power, either as part of a generating station or a local power system such as a nuclear submarine.

- *research reactors* are intended to produce neutrons and/or activate radioactive sources for scientific, medical, engineering, or other research purposes.

- *breeder reactors* are intended to produce nuclear fuels in bulk from more abundant isotopes. The better known fast breeder reactor makes ^{239}Pu (a nuclear fuel) from the naturally very abundant ^{238}U (not a nuclear fuel). Thermal breeder reactors previously tested using ^{232}Th to breed the fissile isotope ^{233}U (thorium fuel cycle) continue to be studied and developed.

While, in principle, all fission reactors can act in all three capacities, in practice the tasks lead to conflicting engineering goals and most reactors have been built with only one of the above tasks in mind. (There are several early counter-examples, such as the Hanford N reactor, now decommissioned). Power reactors generally convert the kinetic energy of fission products into heat, which is used to heat a working fluid and drive a heat engine that generates mechanical or electrical power. The working fluid is usually water with a steam turbine, but some designs use other materials such as gaseous helium. Research reactors produce neutrons that are used in various ways, with the heat of fission being treated as an unavoidable waste product. Breeder reactors are a specialized form of research reactor, with the caveat that the sample being irradiated is usually the fuel itself, a mixture of ^{238}U and ^{235}U. For a more detailed description of the physics and operating principles of critical fission reactors, see nuclear reactor physics. For a description of their social, political, and environmental aspects, see nuclear power.

3.1.7   Fission bombs

One class of nuclear weapon, a *fission bomb* (not to be confused with the *fusion bomb*), otherwise known as an *atomic bomb* or *atom bomb*, is a fission reactor designed to liberate as much energy as possible as rapidly as possible, before the released energy causes the reactor to explode (and the chain reaction to stop). Development of nuclear weapons was the motivation behind early research into nuclear fission: the Manhattan Project of the U.S. military during World War II carried out most of the early scientific work on fission chain reactions, culminating in the Trinity test bomb and the Little Boy and Fat Man bombs that were exploded over the cities Hiroshima, and Nagasaki, Japan in August 1945.

Even the first fission bombs were thousands of times more explosive than a comparable mass of chemical explosive. For example, Little Boy weighed a total of about four tons (of which 60 kg was nuclear fuel) and was 11 feet (3.4 m) long; it also yielded an explosion equivalent to about 15 kilotons of TNT, destroying a large part of the city of Hiroshima. Modern nuclear weapons (which include a thermonuclear *fusion* as well as one or more fission stages) are hundreds of times more energetic for their weight than the first pure fission atomic bombs (see nuclear weapon yield), so that a modern single missile warhead bomb weighing less than 1/8 as much as Little Boy (see for example W88) has a yield of 475,000 tons of TNT, and could bring destruction to about 10 times the city area.

While the fundamental physics of the fission chain reaction in a nuclear weapon is similar to the physics of a controlled nuclear reactor, the two types of device must be engineered quite differently (see nuclear reactor physics). A nuclear bomb is designed to release all its energy at once, while a reactor is designed to generate a steady supply of useful power. While overheating of a reactor can lead to, and has led to, meltdown and steam explosions, the much lower uranium enrichment makes it impossible for a nuclear reactor to explode with the same destructive power as a nuclear weapon. It is also difficult to extract useful power from a nuclear bomb, although at least one rocket propulsion system, Project Orion, was intended to work by exploding fission bombs behind a massively padded and shielded spacecraft.

The strategic importance of nuclear weapons is a major reason why the technology of nuclear fission is politically sensitive. Viable fission bomb designs are, arguably, within the capabilities of many being relatively simple from an engineering viewpoint. However, the difficulty of obtaining fissile nuclear material to realize the designs, is the key to the relative unavailability of nuclear weapons to all but modern industrialized governments with special programs to produce fissile materials (see uranium enrichment and nuclear fuel cycle).

3.2   History

3.2.1   Discovery of nuclear fission

The discovery of nuclear fission occurred in 1938 in the buildings of Kaiser Wilhelm Society for Chemistry, today part of the Free University of Berlin, following nearly five decades of work on the science of radioactivity and the elaboration of new nuclear physics that described the components of atoms. In 1911, Ernest Rutherford proposed a model of the atom in which a very small, dense and positively charged nucleus of protons (the neutron had not yet been discovered) was surrounded by orbiting, negatively charged electrons (the Rutherford model).[13] Niels Bohr improved upon this in 1913 by reconciling the quantum behavior of electrons (the Bohr model). Work by Henri Becquerel, Marie Curie, Pierre Curie, and Rutherford further elaborated that the nucleus, though tightly bound, could undergo different forms of radioactive decay, and thereby transmute into other elements. (For example, by alpha decay: the emission of an alpha particle—two protons and two neutrons bound together into a particle identical to a helium nucleus.)

Some work in nuclear transmutation had been done. In 1917, Rutherford was able to accomplish transmutation of nitrogen into oxygen, using alpha particles directed at nitrogen ^{14}N + α → ^{17}O + p. This was the first observation of a nuclear reaction, that is, a reaction in which particles from one decay are used to transform another atomic nucleus. Eventually, in 1932, a fully artificial nuclear reaction and nuclear transmutation was achieved by Rutherford's colleagues Ernest

Walton and John Cockcroft, who used artificially accelerated protons against lithium-7, to split this nucleus into two alpha particles. The feat was popularly known as "splitting the atom", although it was not the modern nuclear fission reaction later discovered in heavy elements, which is discussed below.[14] Meanwhile, the possibility of *combining* nuclei—nuclear fusion—had been studied in connection with understanding the processes which power stars. The first artificial fusion reaction had been achieved by Mark Oliphant in 1932, using two accelerated deuterium nuclei (each consisting of a single proton bound to a single neutron) to create a helium nucleus.[15]

After English physicist James Chadwick discovered the neutron in 1932,[16] Enrico Fermi and his colleagues in Rome studied the results of bombarding uranium with neutrons in 1934.[17] Fermi concluded that his experiments had created new elements with 93 and 94 protons, which the group dubbed ausonium and hesperium. However, not all were convinced by Fermi's analysis of his results. The German chemist Ida Noddack notably suggested in print in 1934 that instead of creating a new, heavier element 93, that "it is conceivable that the nucleus breaks up into several large fragments."[18][19] However, Noddack's conclusion was not pursued at the time.

After the Fermi publication, Otto Hahn, Lise Meitner, and Fritz Strassmann began performing similar experiments in Berlin. Meitner, an Austrian Jew, lost her citizenship with the "Anschluss", the occupation and annexation of Austria into Nazi Germany in March 1938, but she fled in July 1938 to Sweden and started a correspondence by mail with Hahn in Berlin. By coincidence, her nephew Otto Robert Frisch, also a refugee, was also in Sweden when Meitner received a letter from Hahn dated 19 December describing his chemical proof that some of the product of the bombardment of uranium with neutrons was barium. Hahn suggested a *bursting* of the nucleus, but he was unsure of what the physical basis for the results were. Barium had an atomic mass 40% less than uranium, and no previously known methods of radioactive decay could account for such a large difference in the mass of the nucleus. Frisch was skeptical, but Meitner trusted Hahn's ability as a chemist. Marie Curie had been separating barium from radium for many years, and the techniques were well-known. According to Frisch:

> Was it a mistake? No, said Lise Meitner; Hahn was too good a chemist for that. But how could barium be formed from uranium? No larger fragments than protons or helium nuclei (alpha particles) had ever been chipped away from nuclei, and to chip off a large number not nearly enough energy was available. Nor was it possible that the uranium nucleus could have been cleaved right across. A nucleus was not like a brittle solid that can be cleaved or broken; George Gamow had suggested early on, and Bohr had given good arguments that a nucleus was much more like a liquid drop. Perhaps a drop could divide itself into two smaller drops in a more gradual manner, by first becoming elongated, then constricted, and finally being torn rather than broken in two? We knew that there were strong forces that would resist such a process, just as the surface tension of an ordinary liquid drop tends to resist its division into two smaller ones. But nuclei differed from ordinary drops in one important way: they were electrically charged, and that was known to counteract the surface tension.

> The charge of a uranium nucleus, we found, was indeed large enough to overcome the effect of the surface tension almost completely; so the uranium nucleus might indeed resemble a very wobbly unstable drop, ready to divide itself at the slightest provocation, such as the impact of a single neutron. But there was another problem. After separation, the two drops would be driven apart by their mutual electric repulsion and would acquire high speed and hence a very large energy, about 200 MeV in all; where could that energy come from? ...Lise Meitner... worked out that the two nuclei formed by the division of a uranium nucleus together would be lighter than the original uranium nucleus by about one-fifth the mass of a proton. Now whenever mass disappears energy is created, according to Einstein's formula $E = mc^2$, and one-fifth of a proton mass was just equivalent to 200 MeV. So here was the source for that energy; it all fitted![20]

In short, Meitner and Frisch had correctly interpreted Hahn's results to mean that the nucleus of uranium had split roughly in half. Frisch suggested the process be named "nuclear fission," by analogy to the process of living cell division into two cells, which was then called binary fission. Just as the term nuclear "chain reaction" would later be borrowed from chemistry, so the term "fission" was borrowed from biology.

On 22 December 1938, Hahn and Strassmann sent a manuscript to *Naturwissenschaften* reporting that they had discovered the element barium after bombarding uranium with neutrons.[21] Simultaneously, they communicated these results to Meitner in Sweden. She and Frisch correctly interpreted the results as evidence of nuclear fission.[22] Frisch confirmed this experimentally on 13 January 1939.[23][24] For proving that the barium resulting from his bombardment of uranium with neutrons was the product of nuclear fission, Hahn was awarded the Nobel Prize for Chemistry in 1944 (the sole recipient) "for his discovery of the fission of heavy nuclei". (The award was actually given to Hahn in 1945, as "the Nobel Committee for Chemistry decided that none of the year's nominations met the criteria as outlined in the will of Alfred Nobel." In such cases, the Nobel Foundation's statutes permit that year's prize be reserved until the following year.)[25]

News spread quickly of the new discovery, which was correctly seen as an entirely novel physical effect with great scientific—and potentially practical—possibilities. Meitner's and Frisch's interpretation of the discovery of Hahn and Strassmann crossed the Atlantic Ocean with Niels Bohr, who was to lecture at Princeton University. I.I. Rabi and Willis Lamb, two Columbia University physicists working at Princeton, heard the news and carried it back to Columbia. Rabi said he told Enrico Fermi; Fermi gave credit to Lamb. Bohr soon thereafter went from Princeton to Columbia to see Fermi. Not finding Fermi in his office, Bohr went down to the cyclotron area and found Herbert L. Anderson. Bohr grabbed him by the shoulder and said: "Young man, let me explain to you about something new and exciting in physics."[26] It was clear to a number of scientists at Columbia that they should try to detect the energy released in the nuclear fission of uranium from neutron bombardment. On 25 January 1939, a Columbia University team conducted the first nuclear fission experiment in the United States,[27] which was done in the basement of Pupin Hall; the members of the team were Herbert L. Anderson, Eugene T. Booth, John R. Dunning, Enrico Fermi, G. Norris Glasoe, and Francis G. Slack. The experiment involved placing uranium oxide inside of an ionization chamber and irradiating it with neutrons, and measuring the energy thus released. The results confirmed that fission was occurring and hinted strongly that it was the isotope uranium 235 in particular that was fissioning. The next day, the Fifth Washington Conference on Theoretical Physics began in Washington, D.C. under the joint auspices of the George Washington University and the Carnegie Institution of Washington. There, the news on nuclear fission was spread even further, which fostered many more experimental demonstrations.[28]

During this period the Hungarian physicist Leó Szilárd, who was residing in the United States at the time, realized that the neutron-driven fission of heavy atoms could be used to create a nuclear chain reaction. Such a reaction using neutrons was an idea he had first formulated in 1933, upon reading Rutherford's disparaging remarks about generating power from his team's 1932 experiment using protons to split lithium. However, Szilárd had not been able to achieve a neutron-driven chain reaction with neutron-rich light atoms. In theory, if in a neutron-driven chain reaction the number of secondary neutrons produced was greater than one, then each such reaction could trigger multiple additional reactions, producing an exponentially increasing number of reactions. It was thus a possibility that the fission of uranium could yield vast amounts of energy for civilian or military purposes (i.e., electric power generation or atomic bombs).

Szilard now urged Fermi (in New York) and Frédéric Joliot-Curie (in Paris) to refrain from publishing on the possibility of a chain reaction, lest the Nazi government become aware of the possibilities on the eve of what would later be known as World War II. With some hesitation Fermi agreed to self-censor. But Joliot-Curie did not, and in April 1939 his team in Paris, including Hans von Halban and Lew Kowarski, reported in the journal *Nature* that the number of neutrons emitted with nuclear fission of ^{235}U was then reported at 3.5 per fission.[29] (They later corrected this to 2.6 per fission.) Simultaneous work by Szilard and Walter Zinn confirmed these results. The results suggested the possibility of building nuclear reactors (first called "neutronic reactors" by Szilard and Fermi) and even nuclear bombs. However, much was still unknown about fission and chain reaction systems.

3.2.2 Fission chain reaction realized

"Chain reactions" at that time were a known phenomenon in *chemistry*, but the analogous process in nuclear physics, using neutrons, had been foreseen as early as 1933 by Szilárd, although Szilárd at that time had no idea with what materials the process might be initiated. Szilárd considered that neutrons would be ideal for such a situation, since they lacked an electrostatic charge.

With the news of fission neutrons from uranium fission, Szilárd immediately understood the possibility of a nuclear chain reaction using uranium. In the summer, Fermi and Szilard proposed the idea of a nuclear reactor (pile) to mediate this process. The pile would use natural uranium as fuel. Fermi had shown much earlier that neutrons were far more effectively captured by atoms if they were of low energy (so-called "slow" or "thermal" neutrons), because for quantum reasons it made the atoms look like much larger targets to the neutrons. Thus to slow down the secondary neutrons released by the fissioning uranium nuclei, Fermi and Szilard proposed a graphite "moderator," against which the fast, high-energy secondary neutrons would collide, effectively slowing them down. With enough uranium, and with pure-enough graphite, their "pile" could theoretically sustain a slow-neutron chain reaction. This would result in the production of heat, as well as the creation of radioactive fission products.

In August 1939, Szilard and fellow Hungarian refugees physicists Teller and Wigner thought that the Germans might make use of the fission chain reaction and were spurred to attempt to attract the attention of the United States government to the issue. Towards this, they persuaded German-Jewish refugee Albert Einstein to lend his name to a letter directed to President Franklin Roosevelt. The Einstein–Szilárd letter suggested the possibility of a uranium bomb deliverable by ship, which would destroy "an entire harbor and much of the surrounding countryside." The President received the letter on 11 October 1939 — shortly after World War II began in Europe, but two years before U.S. entry into it. Roosevelt ordered that a scientific committee be authorized for overseeing uranium work and allocated a small sum of money for pile research.

In England, James Chadwick proposed an atomic bomb utilizing natural uranium, based on a paper by Rudolf Peierls with the mass needed for critical state being 30–40 tons. In America, J. Robert Oppenheimer thought that a cube of uranium deuteride 10 cm on a side (about 11 kg of uranium) might "blow itself to hell." In this design it was still thought that a moderator would need to be used for nuclear bomb fission (this turned out not to be the case if the fissile isotope was separated). In December, Werner Heisenberg delivered a report to the German Ministry of War on the possibility of a uranium bomb. Most of these models were still under the assumption that the bombs would be powered by slow neutron reactions—and thus be similar to a reactor undergoing a meltdown.

In Birmingham, England, Frisch teamed up with Peierls, a fellow German-Jewish refugee. They had the idea of using a purified mass of the uranium isotope ^{235}U, which had a cross section just determined, and which was much larger than that of ^{238}U or natural uranium (which is 99.3% the latter isotope). Assuming that the cross section for fast-neutron fission of ^{235}U was the same as for slow neutron fission, they determined that a pure ^{235}U bomb could have a critical mass of only 6 kg instead of tons, and that the resulting explosion would be tremendous. (The amount actually turned out to be 15 kg, although several times this amount was used in the actual uranium (Little Boy) bomb). In February 1940 they delivered the Frisch–Peierls memorandum. Ironically, they were still officially considered "enemy aliens" at the time. Glenn Seaborg, Joseph W. Kennedy, Arthur Wahl and Italian-Jewish refugee Emilio Segrè shortly thereafter discovered ^{239}Pu in the decay products of ^{239}U produced by bombarding ^{238}U with neutrons, and determined it to be a fissile material, like ^{235}U.

The possibility of isolating uranium-235 was technically daunting, because uranium-235 and uranium-238 are chemically identical, and vary in their mass by only the weight of three neutrons. However, if a sufficient quantity of uranium-235 could be isolated, it would allow for a fast neutron fission chain reaction. This would be extremely explosive, a true "atomic bomb." The discovery that plutonium-239 could be produced in a nuclear reactor pointed towards another approach to a fast neutron fission bomb. Both approaches were extremely novel and not yet well understood, and there was considerable scientific skepticism at the idea that they could be developed in a short amount of time.

On June 28, 1941, the Office of Scientific Research and Development was formed in the U.S. to mobilize scientific resources and apply the results of research to national defense. In September, Fermi assembled his first nuclear "pile" or reactor, in an attempt to create a slow neutron-induced chain reaction in uranium, but the experiment failed to achieve criticality, due to lack of proper materials, or not enough of the proper materials which were available.

Producing a fission chain reaction in natural uranium fuel was found to be far from trivial. Early nuclear reactors did not use isotopically enriched uranium, and in consequence they were required to use large quantities of highly purified graphite as neutron moderation materials. Use of ordinary water (as opposed to heavy water) in nuclear reactors requires enriched fuel — the partial separation and relative enrichment of the rare ^{235}U isotope from the far more common ^{238}U isotope. Typically, reactors also require inclusion of extremely chemically pure neutron moderator materials such as deuterium (in heavy water), helium, beryllium, or carbon, the latter usually as graphite. (The high purity for carbon is required because many chemical impurities such as the boron-10 component of natural boron, are very strong neutron absorbers and thus poison the chain reaction and end it prematurely.)

Production of such materials at industrial scale had to be solved for nuclear power generation and weapons production to be accomplished. Up to 1940, the total amount of uranium metal produced in the USA was not more than a few grams, and even this was of doubtful purity; of metallic beryllium not more than a few kilograms; and concentrated deuterium oxide (heavy water) not more than a few kilograms. Finally, carbon had never been produced in quantity with anything like the purity required of a moderator.

The problem of producing large amounts of high purity uranium was solved by Frank Spedding using the thermite or "Ames" process. Ames Laboratory was established in 1942 to produce the large amounts of natural (unenriched) uranium metal that would be necessary for the research to come. The critical nuclear chain-reaction success of the Chicago Pile-1 (December 2, 1942) which used unenriched (natural) uranium, like all of the atomic "piles" which produced the plutonium for the atomic bomb, was also due specifically to Szilard's realization that very pure graphite could be used for the moderator of even natural uranium "piles". In wartime Germany, failure to appreciate the qualities of very pure graphite led to reactor designs dependent on heavy water, which in turn was denied the Germans by Allied attacks in Norway, where heavy water was produced. These difficulties—among many others— prevented the Nazis from building a nuclear reactor capable of criticality during the war, although they never put as much effort as the United States into nuclear research, focusing on other technologies (see German nuclear energy project for more details).

3.2.3 Manhattan Project and beyond

See also: Manhattan Project

In the United States, an all-out effort for making atomic weapons was begun in late 1942. This work was taken over by the U.S. Army Corps of Engineers in 1943, and known as the Manhattan Engineer District. The top-secret Manhattan Project, as it was colloquially known, was led by General Leslie R. Groves. Among the project's dozens of sites were: Hanford Site in Washington state, which had the first industrial-scale nuclear reactors; Oak Ridge, Tennessee, which was primarily concerned with uranium enrichment; and Los Alamos, in New Mexico, which was the scientific hub for research on bomb development and design. Other sites, notably the Berkeley Radiation Laboratory and the Metallurgical Laboratory at the University of Chicago, played important contributing roles. Overall scientific direction of the project was managed by the physicist J. Robert Oppenheimer.

In July 1945, the first atomic bomb, dubbed "Trinity", was detonated in the New Mexico desert. It was fueled by plutonium created at Hanford. In August 1945, two more atomic bombs—"Little Boy", a uranium-235 bomb, and "Fat Man", a plutonium bomb—were used against the Japanese cities of Hiroshima and Nagasaki.

In the years after World War II, many countries were involved in the further development of nuclear fission for the purposes of nuclear reactors and nuclear weapons. The UK opened the first commercial nuclear power plant in 1956. In 2013, there are 437 reactors in 31 countries.

3.2.4 Natural fission chain-reactors on Earth

Criticality in nature is uncommon. At three ore deposits at Oklo in Gabon, sixteen sites (the so-called Oklo Fossil Reactors) have been discovered at which self-sustaining nuclear fission took place approximately 2 billion years ago. Unknown until 1972 (but postulated by Paul Kuroda in 1956[30]), when French physicist Francis Perrin discovered the Oklo Fossil Reactors, it was realized that nature had beaten humans to the punch. Large-scale natural uranium fission chain reactions, moderated by normal water, had occurred far in the past and would not be possible now. This ancient process was able to use normal water as a moderator only because 2 billion years before the present, natural uranium was richer in the shorter-lived fissile isotope ^{235}U (about 3%), than natural uranium available today (which is only 0.7%, and must be enriched to 3% to be usable in light-water reactors).

3.3 See also

- Hybrid fusion/fission

- Cold fission

- Nuclear propulsion

- Photofission

3.4 Notes

[1] M. G. Arora and M. Singh (1994). *Nuclear Chemistry*. Anmol Publications. p. 202. ISBN 81-261-1763-X.

[2] Gopal B. Saha (1 November 2010). *Fundamentals of Nuclear Pharmacy*. Springer. pp. 11–. ISBN 978-1-4419-5860-0.

[3] Петржак, Константин (1989). "Как было открыто спонтанное деление" [How spontaneous fission was discovered]. In Черникова, Вера. *Краткий Миг Торжества — О том, как делаются научные открытия* [*Brief Moment of Triumph — About making scientific discoveries*] (in Russian). Наука. pp. 108–112. ISBN 5-02-007779-8.

[4] S. Vermote, et al. (2008) "Comparative study of the ternary particle emission in 243-Cm (nth,f) and 244-Cm(SF)" in *Dynamical aspects of nuclear fission: proceedings of the 6th International Conference*. J. Kliman, M. G. Itkis, S. Gmuca (eds.). World Scientific Publishing Co. Pte. Ltd. Singapore.

[5] J. Byrne (2011) *Neutrons, Nuclei, and Matter*, Dover Publications, Mineola, NY, p. 259, ISBN 978-0-486-48238-5.

[6] Marion Brünglinghaus. "Nuclear fission". European Nuclear Society. Retrieved 2013-01-04.

[7] Hans A. Bethe (April 1950), "The Hydrogen Bomb", *Bulletin of the Atomic Scientists*, p. 99.

[8] These fission neutrons have a wide energy spectrum, with range from 0 to 14 MeV, with mean of 2 MeV and mode (statistics) of 0.75 Mev. See Byrne, op. cite.

[9] "Nuclear Fission and Fusion, and Nuclear Interactions". National Physical Laboratory. Retrieved 2013-01-04.

[10] L. Bonneau; P. Quentin. "Microscopic calculations of potential energy surfaces: fission and fusion properties" (PDF). Retrieved 2008-07-28.

[11] By R.D. Madan and Satya Prakash - *Modern Inorganic Chemistry*

[12] "Frequently Asked Questions #1". Radiation Effects Research Foundation. Retrieved September 18, 2007.

[13] E. Rutherford (1911). "The scattering of α and β particles by matter and the structure of the atom" (PDF). *Philosophical Magazine* **21** (4): 669–688. Bibcode:2012PMag...92..379R. doi:10.1080/14786435.2011.617037.

[14] "Cockcroft and Walton split lithium with high energy protons April 1932". Outreach.phy.cam.ac.uk. 1932-04-14. Retrieved 2013-01-04.

[15] "Sir Mark Oliphant (1901–2000)" (PDF). University of Adelaide. Retrieved 5 October 2013.

[16] Chadwick announced his initial findings in: J. Chadwick (1932). "Possible Existence of a Neutron" (PDF). *Nature* **129** (3252): 312. Bibcode:1932Natur.129Q.312C. doi:10.1038/129312a0. Subsequently he communicated his findings in more detail in: Chadwick, J. (1932). "The existence of a neutron". *Proceedings of the Royal Society A* **136** (830): 692–708. Bibcode:1932RSPSA.136..692C. doi:10.1098/rspa.1932.0112.; and Chadwick, J. (1933). "The Bakerian Lecture: The neutron". *Proceedings of the Royal Society A* **142** (846): 1–25. Bibcode:1933RSPSA.142....1C. doi:10.1098/rspa.1933.0152.

[17] E. Fermi, E. Amaldi, O. D'Agostino, F. Rasetti, and E. Segrè (1934) "Radioattività provocata da bombardamento di neutroni III," *La Ricerca Scientifica*, vol. 5, no. 1, pages 452–453.

[18] Ida Noddack (1934). "Über das Element 93". *Zeitschrift für Angewandte Chemie* **47** (37): 653. doi:10.1002/ange.19340473707.

[19] Tacke, Ida Eva. Astr.ua.edu. Retrieved on 2010-12-24.

[20] Bob Weintraub. *Lise Meitner (1878–1968): Protactinium, Fission, and Meitnerium*. Retrieved on June 8, 2009.

[21] O. Hahn and F. Strassmann (1939). "Über den Nachweis und das Verhalten der bei der Bestrahlung des Urans mittels Neutronen entstehenden Erdalkalimetalle ("On the detection and characteristics of the alkaline earth metals formed by irradiation of uranium with neutrons")". *Naturwissenschaften* **27** (1): 11–15. Bibcode:1939NW.....27...11H. doi:10.1007/BF01488241.. The authors were identified as being at the Kaiser-Wilhelm-Institut für Chemie, Berlin-Dahlem. Received 22 December 1938.

[22] L. Meitner and O. R. Frisch (1939). "Disintegration of Uranium by Neutrons: a New Type of Nuclear Reaction". *Nature* **143** (3615): 239. Bibcode:1939Natur.143..239M. doi:10.1038/143239a0.. The paper is dated 16 January 1939. Meitner is identified as being at the Physical Institute, Academy of Sciences, Stockholm. Frisch is identified as being at the Institute of Theoretical Physics, University of Copenhagen.

[23] O. R. Frisch (1939). "Physical Evidence for the Division of Heavy Nuclei under Neutron Bombardment". *Nature* **143** (3616): 276. Bibcode:1939Natur.143..276F. doi:10.1038/143276a0.

[24] "Physical Evidence for the Division of Heavy Nuclei under Neutron Bombardment". 17 January 1939. Archived from the original on 2008-01-08. The experiment for this letter to the editor was conducted on 13 January 1939; see Richard Rhodes (1986) *The Making of the Atomic Bomb*, Simon and Schuster. pp. 263 and 268, ISBN 0-671-44133-7.

[25] "The Nobel Prize in Chemistry 1944". Nobelprize.org. Retrieved 2008-10-06.

[26] Richard Rhodes. (1986) *The Making of the Atomic Bomb*, Simon and Schuster, p. 268, ISBN 0-671-44133-7.

[27] H. L. Anderson, E. T. Booth, J. R. Dunning, E. Fermi, G. N. Glasoe, and F. G. Slack (1939). "The Fission of Uranium". *Physical Review* **55** (5): 511. Bibcode:1939PhRv...55..511A. doi:10.1103/PhysRev.55.511.2.

[28] Richard Rhodes (1986). *The Making of the Atomic Bomb*, Simon and Schuster, pp. 267–270, ISBN 0-671-44133-7.

[29] H. Von Halban; F. Joliot and L. Kowarski (1939). "Number of Neutrons Liberated in the Nuclear Fission of Uranium". *Nature* **143** (3625): 680. Bibcode:1939Natur.143..680V. doi:10.1038/143680a0.

[30] P. K. Kuroda (1956). "On the Nuclear Physical Stability of the Uranium Minerals" (PDF). *The Journal of Chemical Physics* **25** (4): 781. Bibcode:1956JChPh..25..781K. doi:10.1063/1.1743058.

3.5   References

- *DOE Fundamentals Handbook: Nuclear Physics and Reactor Theory Volume 1* (PDF). U.S. Department of Energy. January 1993. Retrieved 2012-01-03.

- *DOE Fundamentals Handbook: Nuclear Physics and Reactor Theory Volume 2* (PDF). U.S. Department of Energy. January 1993. Retrieved 2012-01-03.

3.6   External links

- The Effects of Nuclear Weapons

- Annotated bibliography for nuclear fission from the Alsos Digital Library

- The Discovery of Nuclear Fission Historical account complete with audio and teacher's guides from the American Institute of Physics History Center

- atomicarchive.com Nuclear Fission Explained

- Nuclear Files.org What is Nuclear Fission?

- Nuclear Fission Animation

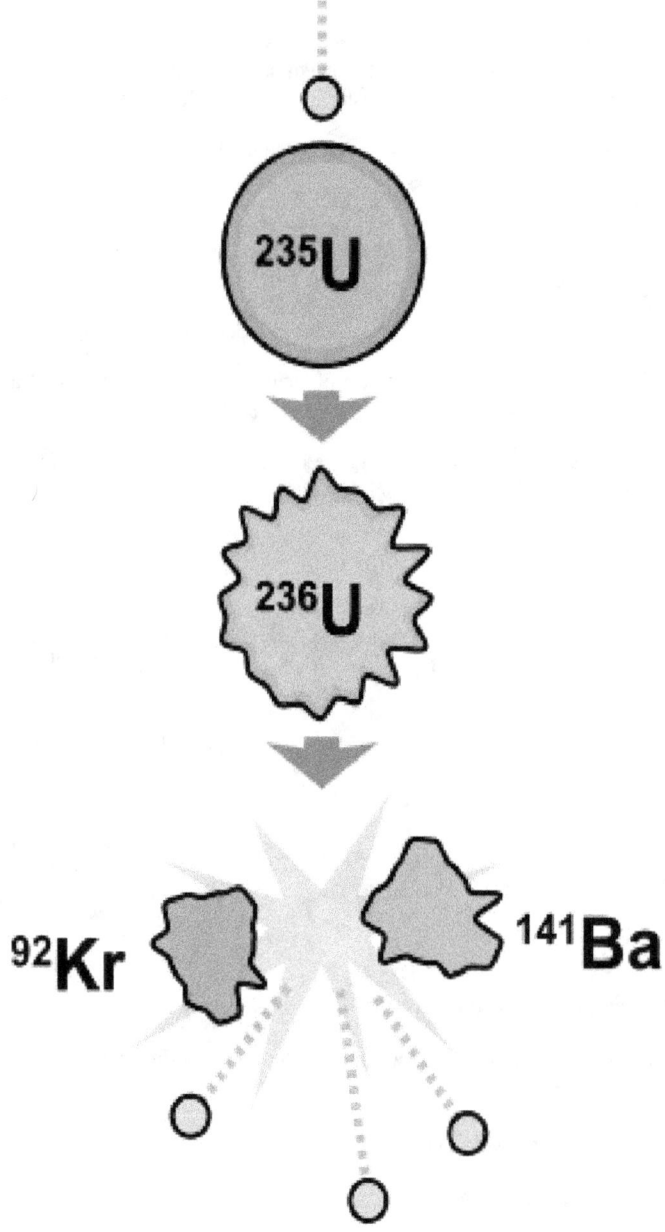

An induced fission reaction. A neutron is absorbed by a uranium-235 nucleus, turning it briefly into an excited uranium-236 nucleus, with the excitation energy provided by the kinetic energy of the neutron plus the forces that bind the neutron. The uranium-236, in turn, splits into fast-moving lighter elements (fission products) and releases three free neutrons. At the same time, one or more "prompt gamma rays" (not shown) are produced, as well.

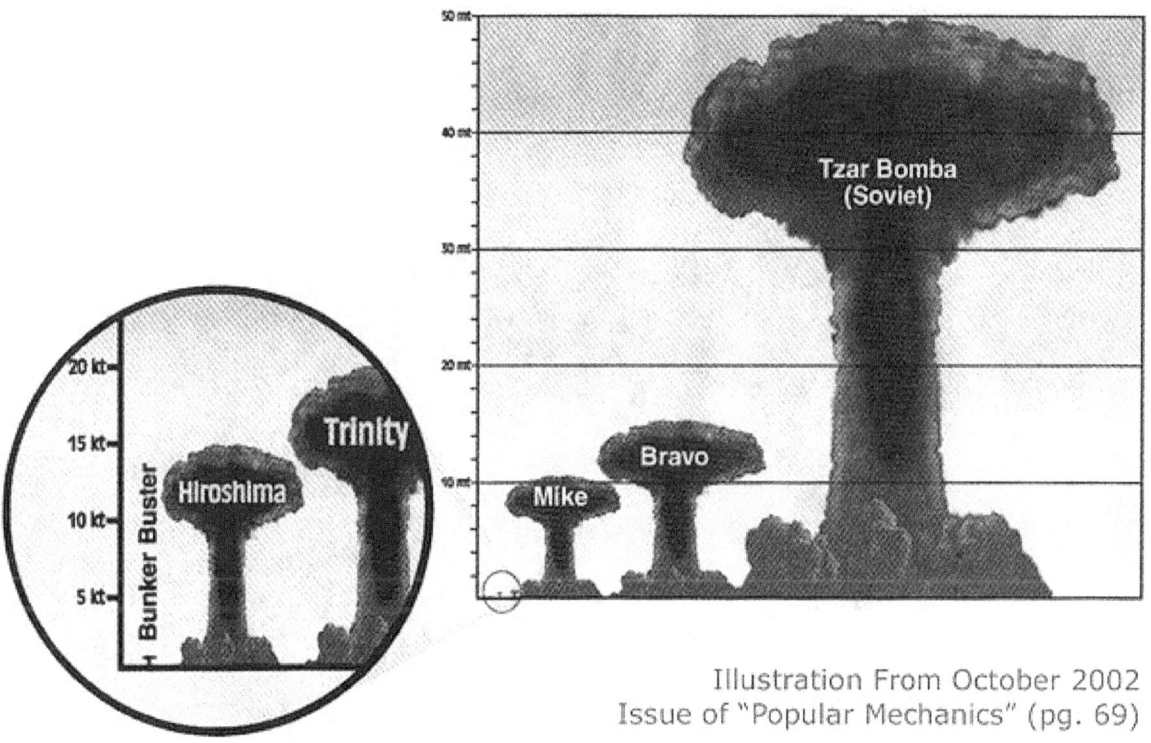

Illustration From October 2002
Issue of "Popular Mechanics" (pg. 69)

The mushroom cloud produced by Tsar Bomba, currently the largest man-made nuclear device detonated in history, next to other mushroom clouds of various nuclear devices.

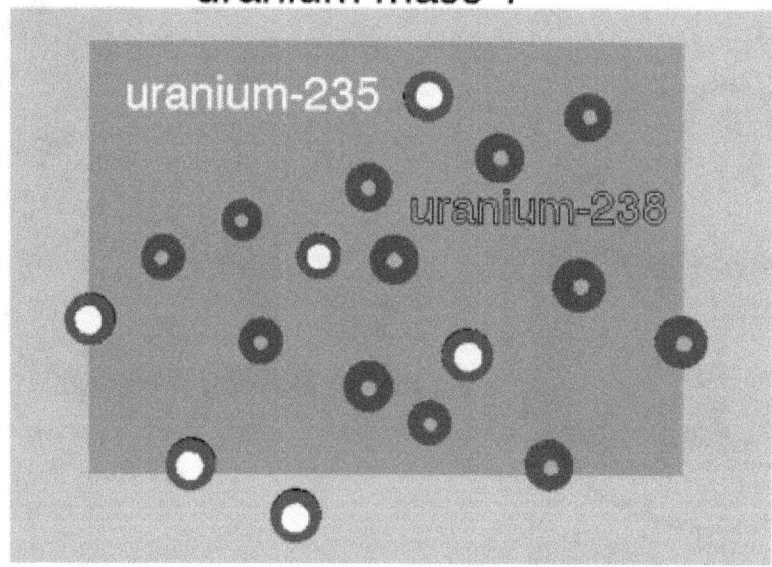

A visual representation of an induced nuclear fission event where a slow-moving neutron is absorbed by the nucleus of a uranium-235 atom, which fissions into two fast-moving lighter elements (fission products) and additional neutrons. Most of the energy released is in the form of the kinetic velocities of the fission products and the neutrons.

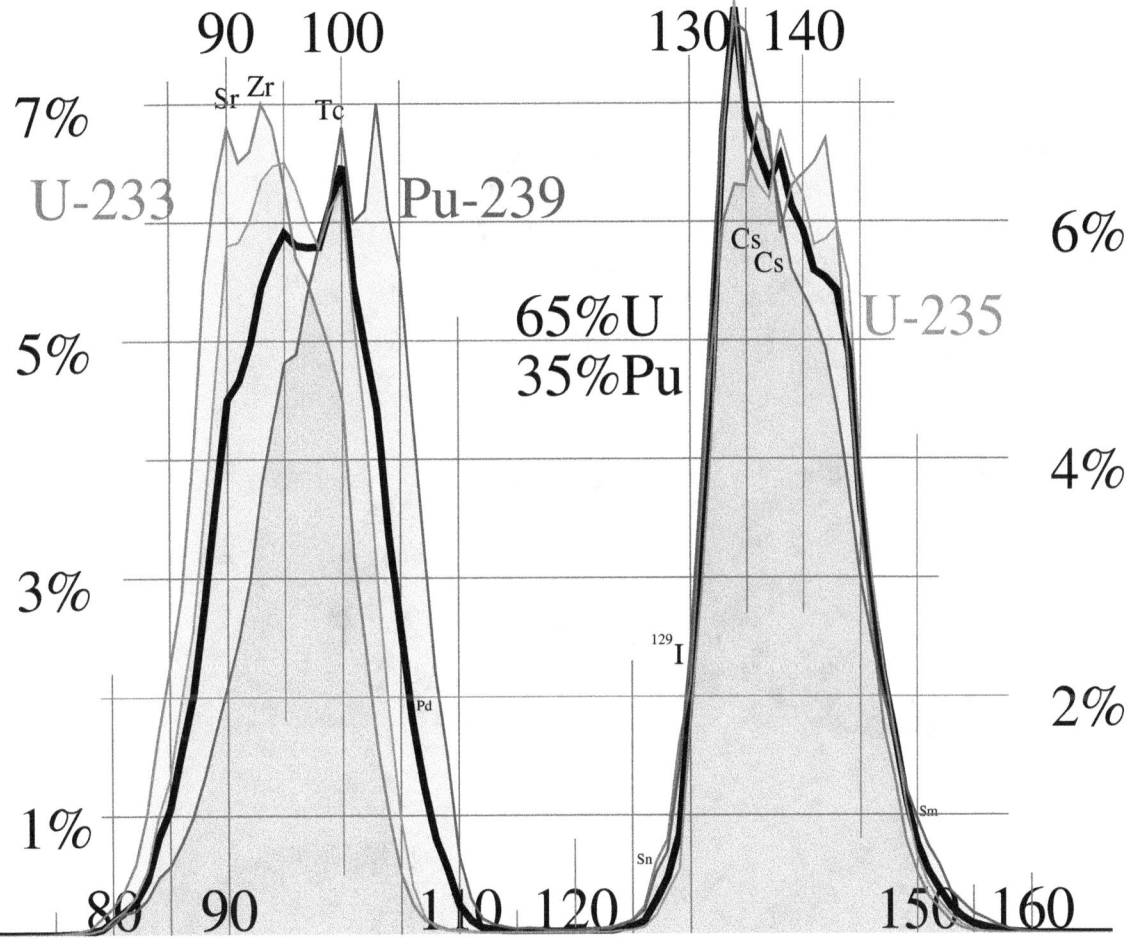

Fission product yields by mass for thermal neutron fission of U-235, Pu-239, a combination of the two typical of current nuclear power reactors, and U-233 used in the thorium cycle.

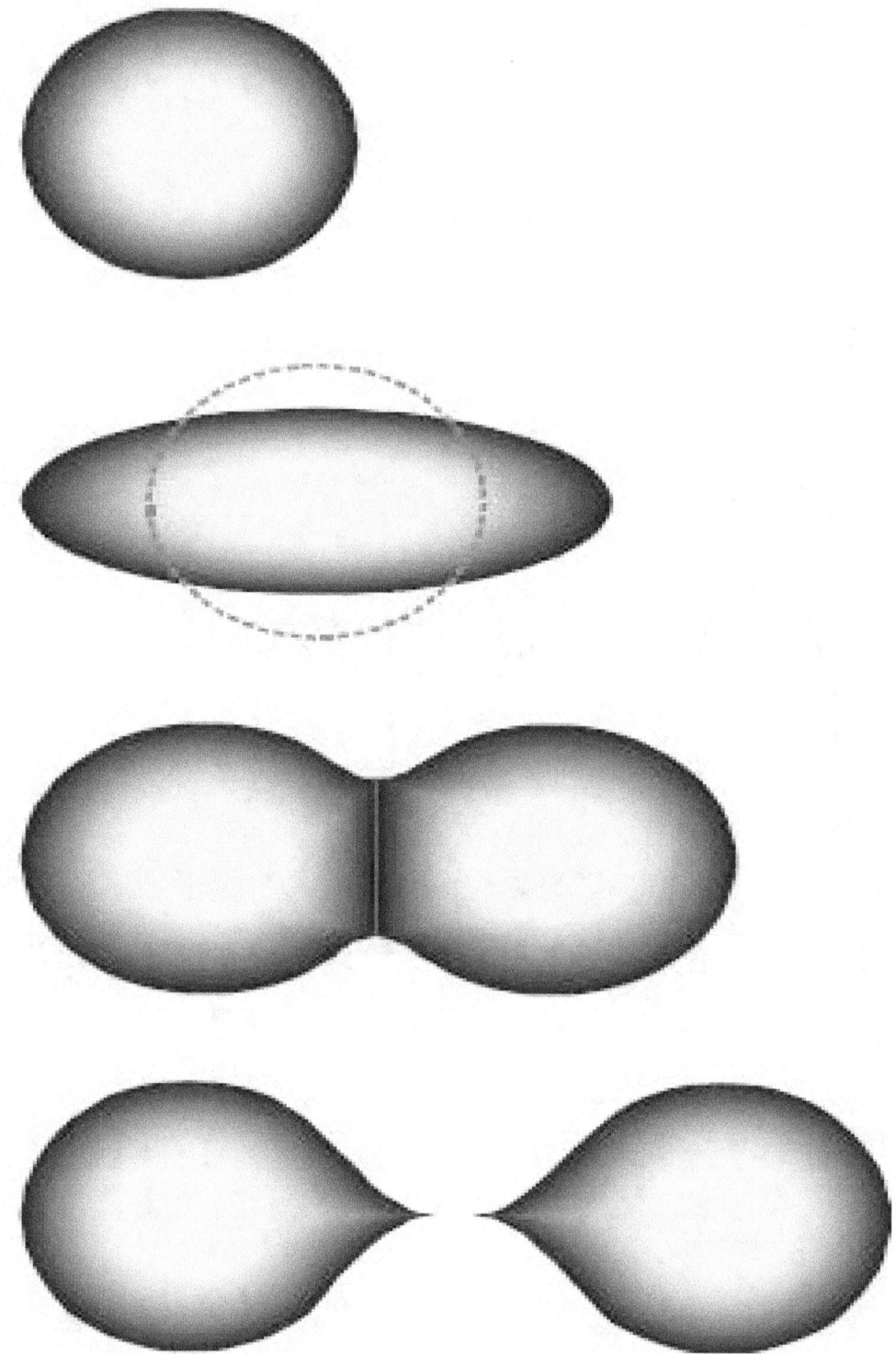

The stages of binary fission in a liquid drop model. Energy input deforms the nucleus into a fat "cigar" shape, then a "peanut" shape, followed by binary fission as the two lobes exceed the short-range nuclear force attraction distance, then are pushed apart and away by their electrical charge. In the liquid drop model, the two fission fragments are predicted to be the same size. The nuclear shell model allows for them to differ in size, as usually experimentally observed.

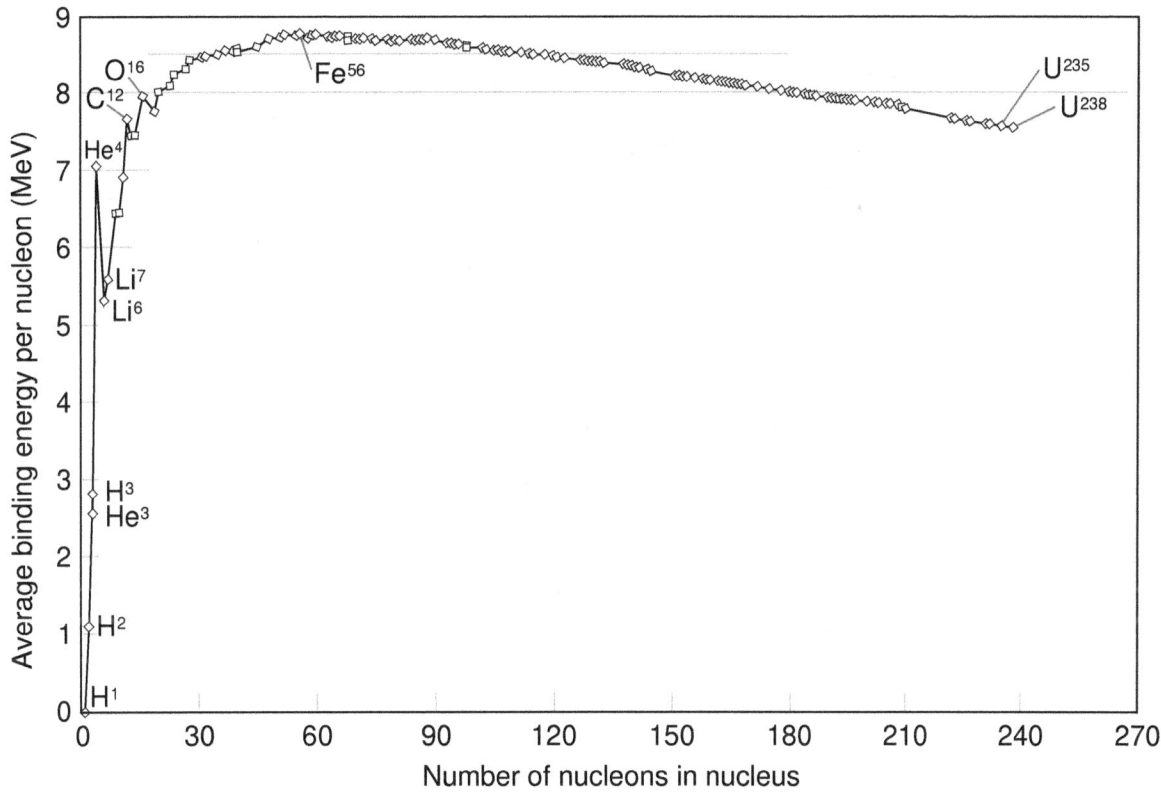

The "curve of binding energy": A graph of binding energy per nucleon of common isotopes.

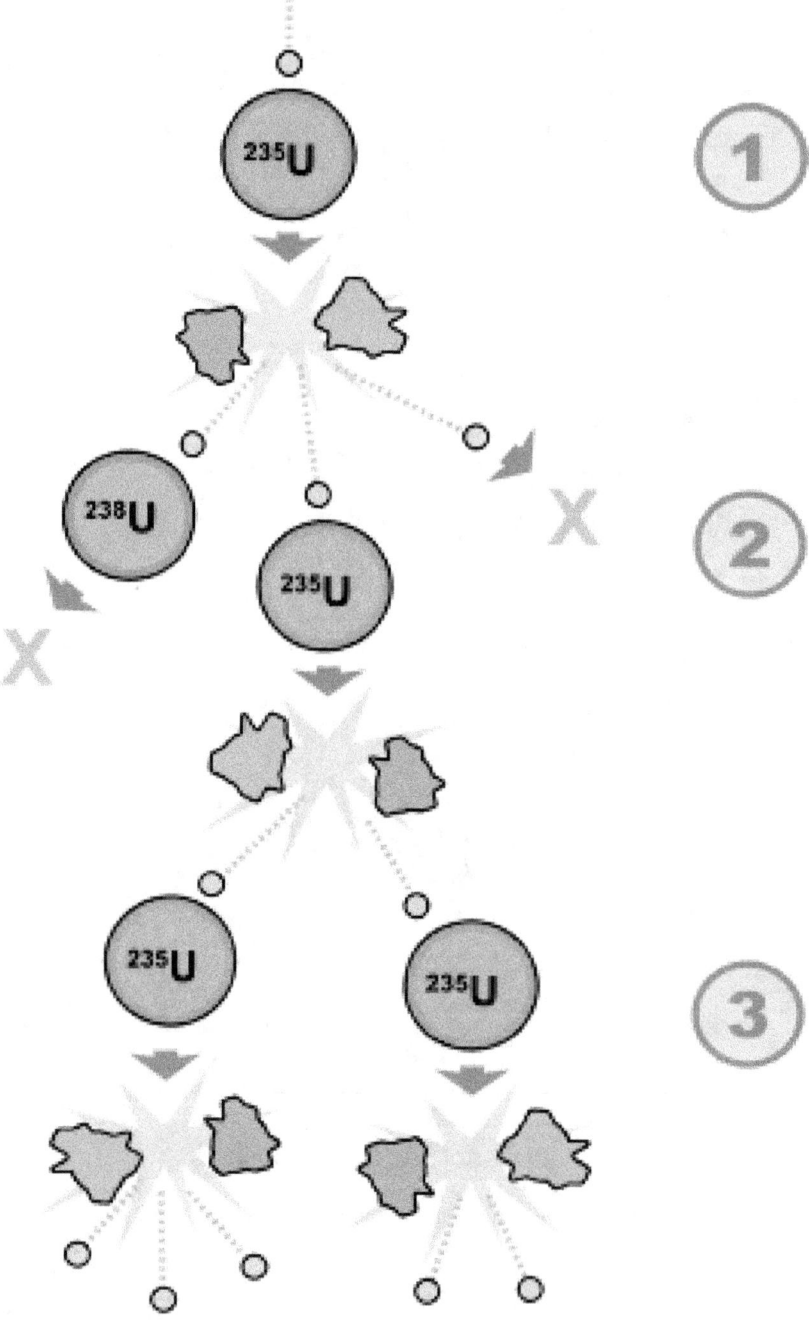

A schematic nuclear fission chain reaction. 1. A uranium-235 atom absorbs a neutron and fissions into two new atoms (fission fragments), releasing three new neutrons and some binding energy. 2. One of those neutrons is absorbed by an atom of uranium-238 and does not continue the reaction. Another neutron is simply lost and does not collide with anything, also not continuing the reaction. However,the one neutron does collide with an atom of uranium-235, which then fissions and releases two neutrons and some binding energy. 3. Both of those neutrons collide with uranium-235 atoms, each of which fissions and releases between one and three neutrons, which can then continue the reaction.

The cooling towers of the Philippsburg Nuclear Power Plant, in Germany.

The mushroom cloud of the atom bomb dropped on Nagasaki, Japan in 1945 rose some 18 kilometres (11 mi) above the bomb's hypocenter. The bomb killed at least 60,000 people.[12]

The experimental apparatus with which Otto Hahn and Fritz Strassmann discovered nuclear fission in 1938

German stamp honoring Otto Hahn and his discovery of nuclear fission (1979)

Drawing of the first artificial reactor, Chicago Pile-1.

Chapter 4

Nucleosynthesis

For the song by Vangelis, see Albedo 0.39.

Nucleosynthesis is the process that creates new atomic nuclei from pre-existing nucleons, primarily protons and neutrons. The first nuclei were formed about three minutes after the Big Bang, through the process called Big Bang nucleosynthesis. It was then that hydrogen and helium formed to become the content of the first stars, and this primeval process is responsible for the present hydrogen/helium ratio of the cosmos.

With the formation of stars, heavier nuclei were created from hydrogen and helium by stellar nucleosynthesis, a process that continues today. Some of these elements, particularly those lighter than iron, continue to be delivered to the interstellar medium when low mass stars eject their outer envelope before they collapse to form white dwarfs. The remains of their ejected mass form the planetary nebulae observable throughout our galaxy.

Supernova nucleosynthesis within exploding stars by fusing carbon and oxygen is responsible for the abundances of elements between magnesium (atomic number 12) and nickel (atomic number 28).[1] Supernova nucleosynthesis is also thought to be responsible for the creation of rarer elements heavier than iron and nickel, in the last few seconds of a type II supernova event. The synthesis of these heavier elements absorbs energy (endothermic) as they are created, from the energy produced during the supernova explosion. Some of those elements are created from the absorption of multiple neutrons (the R process) in the period of a few seconds during the explosion. The elements formed in supernovas include the heaviest elements known, such as the long-lived elements uranium and thorium.

Cosmic ray spallation, caused when cosmic rays impact the interstellar medium and fragment larger atomic species, is a significant source of the lighter nuclei, particularly ^{3}He, ^{9}Be and ^{10,11}B, that are not created by stellar nucleosynthesis.

In addition to the fusion processes responsible for the growing abundances of elements in the universe, a few minor natural processes continue to produce very small numbers of new nuclides on Earth. These nuclides contribute little to their abundances, but may account for the presence of specific new nuclei. These nuclides are produced via radiogenesis (decay) of long-lived, heavy, primordial radionuclides such as uranium and thorium. Cosmic ray bombardment of elements on Earth also contribute to the presence of rare, short-lived atomic species called cosmogenic nuclides.

4.1 Timeline

It is thought that the primordial nucleons themselves were formed from the quark–gluon plasma during the Big Bang as it cooled below two trillion degrees. A few minutes afterward, starting with only protons and neutrons, nuclei up to lithium and beryllium (both with mass number 7) were formed, but the abundances of other elements dropped sharply with growing atomic mass. Some boron may have been formed at this time, but the process stopped before significant carbon could be formed, as this element requires a far higher product of helium density and time than were present in the short nucleosynthesis period of the Big Bang. That fusion process essentially shut down at about 20 minutes, due to drops in temperature and density as the universe continued to expand. This first process, Big Bang nucleosynthesis, was

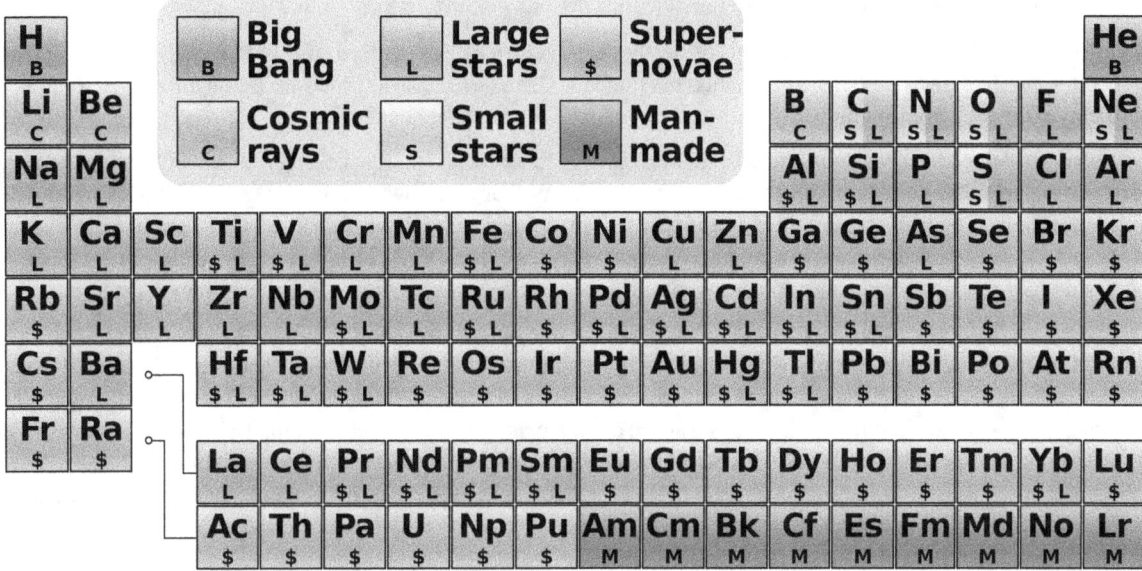

Periodic table showing the cosmogenic origin of each element. Elements from carbon up to sulfur may be made in small stars by the alpha process. Elements beyond iron are made in large stars with slow neutron capture (s-process), followed by expulsion to space in gas ejections (see planetary nebulae). Elements heavier than iron may be made in supernovae after the r-process, involving a dense burst of neutrons and rapid capture by the element.

the first type of nucleogenesis to occur in the universe.

The subsequent nucleosynthesis of the heavier elements requires the extreme temperatures and pressures found within stars and supernovas. These processes began as hydrogen and helium from the Big Bang collapsed into the first stars at 500 million years. Star formation has occurred continuously in the galaxy since that time. The elements found on Earth, the so-called primordial elements, were created prior to Earth's formation by stellar nucleosynthesis and by supernova nucleosynthesis. They range in atomic numbers from Z=6 (carbon) to Z=94 (plutonium). Synthesis of these elements occurred either by nuclear fusion (including both rapid and slow multiple neutron capture) or to a lesser degree by nuclear fission followed by beta decay.

A star gains heavier elements by combining its lighter nuclei, hydrogen, deuterium, beryllium, lithium, and boron, which were found in the initial composition of the interstellar medium and hence the star. Interstellar gas therefore contains declining abundances of these light elements, which are present only by virtue of their nucleosynthesis during the Big Bang. Larger quantities of these lighter elements in the present universe are therefore thought to have been restored through billions of years of cosmic ray (mostly high-energy proton) mediated breakup of heavier elements in interstellar gas and dust. The fragments of these cosmic-ray collisions include the light elements Li, Be and B.

4.2   History of nucleosynthesis theory

The first ideas on nucleosynthesis were simply that the chemical elements were created at the beginning of the universe, but no rational physical scenario for this could be identified. Gradually it became clear that hydrogen and helium are much more abundant than any of the other elements. All the rest constitute less than 2% of the mass of the Solar System, and of other star systems as well. At the same time it was clear that oxygen and carbon were the next two most common elements, and also that there was a general trend toward high abundance of the light elements, especially those composed of whole numbers of helium-4 nuclei.

Arthur Stanley Eddington first suggested in 1920, that stars obtain their energy by fusing hydrogen into helium and raised the possibility that the heavier elements may also form in stars.[2][3] This idea was not generally accepted, as the nuclear mechanism was not understood. In the years immediately before World War II, Hans Bethe first elucidated those nuclear mechanisms by which hydrogen is fused into helium.

Fred Hoyle's original work on nucleosynthesis of heavier elements in stars, occurred just after World War II.[4] His work explained the production of all heavier elements, starting from hydrogen. Hoyle proposed that hydrogen is continuously created in the universe from vacuum and energy, without need for universal beginning.

Hoyle's work explained how the abundances of the elements increased with time as the galaxy aged. Subsequently, Hoyle's picture was expanded during the 1960s by contributions from William A. Fowler, Alastair G. W. Cameron, and Donald D. Clayton, followed by many others. In the seminal 1957 review paper by E. M. Burbidge, G. R. Burbidge, Fowler and Hoyle (see Ref. list) is a well-known summary of the state of the field in 1957. That paper defined new processes for the transformation of one heavy nucleus into others within stars, processes that could be documented by astronomers.

The Big Bang itself had been proposed in 1931, long before this period, by Georges Lemaître, a Belgian physicist, who suggested that the evident expansion of the Universe in time required that the Universe, if contracted backwards in time, would continue to do so until it could contract no further. This would bring all the mass of the Universe to a single point, a "primeval atom", to a state before which time and space did not exist. Hoyle later gave Lemaître's model the derisive term of Big Bang, not realizing that Lemaître's model was needed to explain the existence of deuterium and nuclides between helium and carbon, as well as the fundamentally high amount of helium present, not only in stars but also in interstellar space. As it happened, both Lemaître and Hoyle's models of nucleosynthesis would be needed to explain the elemental abundances in the universe.

The goal of the theory of nucleosynthesis is to explain the vastly differing abundances of the chemical elements and their several isotopes from the perspective of natural processes. The primary stimulus to the development of this theory was the shape of a plot of the abundances versus the atomic number of the elements. Those abundances, when plotted on a graph as a function of atomic number, have a jagged sawtooth structure that varies by factors up to ten million. A very influential stimulus to nucleosynthesis research was an abundance table created by Hans Suess and Harold Urey that was based on the unfractionated abundances of the non-volatile elements found within unevolved meteorites.[5] Such a graph of the abundances is displayed on a logarithmic scale below, where the dramatically jagged structure is visually suppressed by the many powers of ten spanned in the vertical scale of this graph. See *Handbook of Isotopes in the Cosmos* for more data and discussion of abundances of the isotopes.[6]

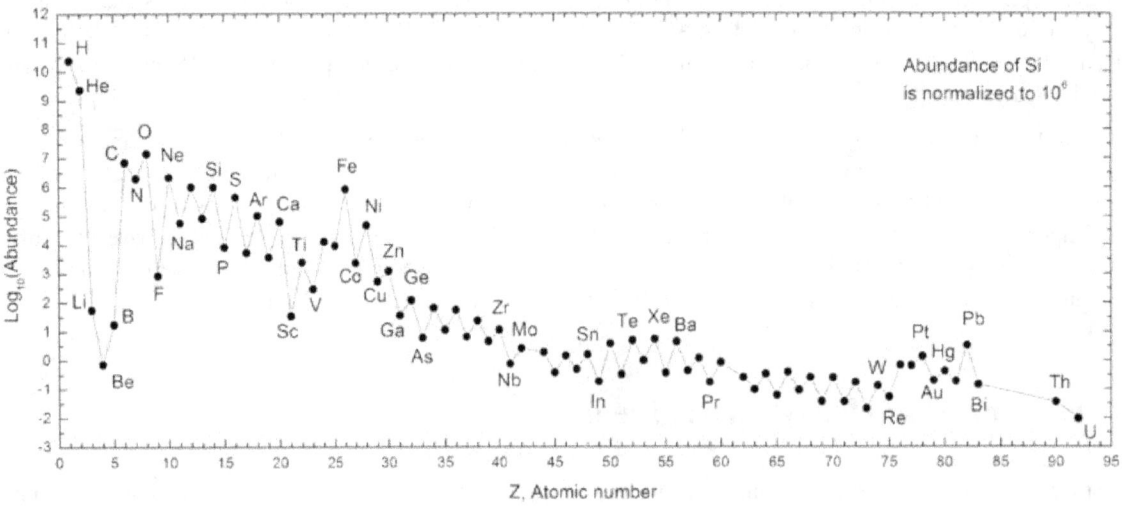

Abundances of the chemical elements in the Solar System. Hydrogen and helium are most common, residuals within the paradigm of the Big Bang.[7] The next three elements (Li, Be, B) are rare because they are poorly synthesized in the Big Bang and also in stars. The two general trends in the remaining stellar-produced elements are: (1) an alternation of abundance of elements according to whether they have even or odd atomic numbers, and (2) a general decrease in abundance, as elements become heavier. Within this trend is a peak at abundances of iron and nickel, which is especially visible on a logarithmic graph spanning fewer powers of ten, say between logA=2 (A=100) and logA=6 (A=1,000,000).

4.3   Processes

There are a number of astrophysical processes which are believed to be responsible for nucleosynthesis. The majority of these occur in shells within stars, and the chain of those nuclear fusion processes are known as hydrogen burning (via the proton-proton chain or the CNO cycle), helium burning, carbon burning, neon burning, oxygen burning and silicon burning. These processes are able to create elements up to and including iron and nickel. This is the region of nucleosynthesis within which the isotopes with the highest binding energy per nucleon are created. Heavier elements can be assembled within stars by a neutron capture process known as the s-process or in explosive environments, such as supernovae, by a number of other processes. Some of those others include the r-process, which involves rapid neutron captures, the rp-process, and the p-process (sometimes known as the gamma process), which results in the photodisintegration of existing nuclei.

4.4   The major types of nucleosynthesis

4.4.1   Big Bang nucleosynthesis

Main article: Big Bang nucleosynthesis

Big Bang nucleosynthesis occurred within the first three minutes of the beginning of the universe and is responsible for much of the abundance of ^{1}H (protium), ^{2}H (D, deuterium), ^{3}He (helium-3), and ^{4}He (helium-4). Although ^{4}He continues to be produced by stellar fusion and alpha decays and trace amounts of ^{1}H continue to be produced by spallation and certain types of radioactive decay, most of the mass of the isotopes in the universe are thought to have been produced in the Big Bang. The nuclei of these elements, along with some ^{7}Li and ^{7}Be are considered to have been formed between 100 and 300 seconds after the Big Bang when the primordial quark–gluon plasma froze out to form protons and neutrons. Because of the very short period in which nucleosynthesis occurred before it was stopped by expansion and cooling (about 20 minutes), no elements heavier than beryllium (or possibly boron) could be formed. Elements formed during this time were in the plasma state, and did not cool to the state of neutral atoms until much later.

$$n^0 \longrightarrow p^+ + e^- + \bar{\nu}_e \qquad\qquad p^+ + n^0 \longrightarrow {}^2_1D + \gamma$$

$$\textstyle {}^2_1D + p^+ \longrightarrow {}^3_2He + \gamma \qquad\qquad {}^2_1D + {}^2_1D \longrightarrow {}^3_2He + n^0$$

$$\textstyle {}^2_1D + {}^2_1D \longrightarrow {}^3_1T + p^+ \qquad\qquad {}^3_1T + {}^2_1D \longrightarrow {}^4_2He + n^0$$

$$\textstyle {}^3_1T + {}^4_2He \longrightarrow {}^7_3Li + \gamma \qquad\qquad {}^3_2He + n^0 \longrightarrow {}^3_1T + p^+$$

$$\textstyle {}^3_2He + {}^2_1D \longrightarrow {}^4_2He + p^+ \qquad\qquad {}^3_2He + {}^4_2He \longrightarrow {}^7_4Be + \gamma$$

$$\textstyle {}^7_3Li + p^+ \longrightarrow {}^4_2He + {}^4_2He \qquad\qquad {}^7_4Be + n^0 \longrightarrow {}^7_3Li + p^+$$

Chief nuclear reactions responsible for the relative abundances of light atomic nuclei observed throughout the universe.

4.4.2   Stellar nucleosynthesis

Main articles: Stellar nucleosynthesis, Proton-proton chain, Triple-alpha process, CNO cycle, s-process, p-process and photodisintegration

Stellar nucleosynthesis is the nuclear process by which new nuclei are produced. It occurs in stars during stellar evolution. It is responsible for the galactic abundances of elements from carbon to iron. Stars are thermonuclear furnaces in which H and He are fused into heavier nuclei by increasingly high temperatures as the composition of the core evolves.[8] Of particular importance is carbon, because its formation from He is a bottleneck in the entire process. Carbon is produced by the triple-alpha process in all stars. Carbon is also the main element that causes the release of free neutrons within stars, giving rise to the s-process, in which the slow absorption of neutrons converts iron into elements heavier than iron and nickel.[9]

The products of stellar nucleosynthesis are generally dispersed into the interstellar gas through mass loss episodes and the stellar winds of low mass stars. The mass loss events can be witnessed today in the planetary nebulae phase of low-mass star evolution, and the explosive ending of stars, called supernovae, of those with more than eight times the mass of the Sun.

The first direct proof that nucleosynthesis occurs in stars was the astronomical observation that interstellar gas has become enriched with heavy elements as time passed. As a result, stars that were born from it late in the galaxy, formed with much higher initial heavy element abundances than those that had formed earlier. The detection of technetium in the atmosphere of a red giant star in 1952,[10] by spectroscopy, provided the first evidence of nuclear activity within stars. Because technetium is radioactive, with a half-life much less than the age of the star, its abundance must reflect its recent creation within that star. Equally convincing evidence of the stellar origin of heavy elements, is the large overabundances of specific stable elements found in stellar atmospheres of asymptotic giant branch stars. Observation of barium abundances some 20-50 times greater than found in unevolved stars is evidence of the operation of the s-process within such stars. Many modern proofs of stellar nucleosynthesis are provided by the isotopic compositions of stardust, solid grains that have condensed from the gases of individual stars and which have been extracted from meteorites. Stardust is one component of cosmic dust, and is frequently called presolar grains. The measured isotopic compositions in stardust grains demonstrate many aspects of nucleosynthesis within the stars from which the grains condensed during the star's late-life mass-loss episodes.[11]

4.4.3 Explosive nucleosynthesis

Main articles: r-process, rp-process and Supernova nucleosynthesis

Supernova nucleosynthesis occurs in the energetic environment in supernovae, in which the elements between silicon and nickel are synthesized in quasiequilibrium[12] established during fast fusion that attaches by reciprocating balanced nuclear reactions to ^{28}Si. Quasiequilibrium can be thought of as *almost equilibrium* except for a high abundance of the ^{28}Si nuclei in the feverishly burning mix. This concept[13] was the most important discovery in nucleosynthesis theory of the intermediate-mass elements since Hoyle's 1954 paper because it provided an overarching understanding of the abundant and chemically important elements between silicon (A=28) and nickel (A=60). It replaced the incorrect although much cited alpha process of the B2FH paper, which inadvertently obscured Hoyle's better 1954 theory.[14] Further nucleosynthesis processes can occur, in particular the r-process (rapid process) described by the B2FH paper and first calculated by Seeger, Fowler and Clayton,[15] in which the most neutron-rich isotopes of elements heavier than nickel are produced by rapid absorption of free neutrons. The creation of free neutrons by electron capture during the rapid compression of the supernova core along with assembly of some neutron-rich seed nuclei makes the r-process a *primary process*, and one that can occur even in a star of pure H and He. This is in contrast to the B2FH designation of the process as a *secondary process*. This promising scenario, though generally supported by supernova experts, has yet to achieve a totally satisfactory calculation of r-process abundances. The primary r-process has been confirmed by astronomers who have observed old stars born when galactic metallicity was still small, that nonetheless contain their complement of r-process nuclei; thereby demonstrating that the metallicity is a product of an internal process. The r-process is responsible for our natural cohort of radioactive elements, such as uranium and thorium, as well as the most neutron-rich isotopes of each heavy element.

The rp-process (rapid proton) involves the rapid absorption of free protons as well as neutrons, but its role and its existence are less certain.

Explosive nucleosynthesis occurs too rapidly for radioactive decay to decrease the number of neutrons, so that many abundant isotopes with equal and even numbers of protons and neutrons are synthesized by the silicon quasiequilibrium process.[16] During this process, the burning of oxygen and silicon fuses nuclei that themselves have equal numbers of protons and neutrons to produce nuclides which consist of whole numbers of helium nuclei, up to 15 (representing ^{60}Ni). Such multiple-alpha-particle nuclides are totally stable up to ^{40}Ca (made of 10 helium nuclei), but heavier nuclei with equal and even numbers of protons and neutrons are tightly bound but unstable. The quasiequilibrium produces radioactive isobars ^{44}Ti, ^{48}Cr, ^{52}Fe, and ^{56}Ni, which (except ^{44}Ti) are created in abundance but decay after the explosion and leave the most stable isotope of the corresponding element at the same atomic weight. The most abundant and extant isotopes of elements produced in this way are ^{48}Ti, ^{52}Cr, and ^{56}Fe. These decays are accompanied by the emission of gamma-rays (radiation from the nucleus), whose spectroscopic lines can be used to identify the isotope created by the decay. The detection of these emission lines were an important early product of gamma-ray astronomy.[17]

The most convincing proof of explosive nucleosynthesis in supernovae occurred in 1987 when those gamma-ray lines were detected emerging from supernova 1987A. Gamma ray lines identifying ^{56}Co and ^{57}Co nuclei, whose radioactive half-lives limit their age to about a year, proved that they were created by their radioactive cobalt parents. This nuclear astronomy observation was predicted in 1969[18] as a way to confirm explosive nucleosynthesis of the elements, and that prediction played an important role in the planning for NASA's Compton Gamma-Ray Observatory.

Other proofs of explosive nucleosynthesis are found within the stardust grains that condensed within the interiors of supernovae as they expanded and cooled. Stardust grains are one component of cosmic dust. In particular, radioactive ^{44}Ti was measured to be very abundant within supernova stardust grains at the time they condensed during the supernova expansion.[19] This confirmed a 1975 prediction of the identification of supernova stardust (SUNOCONs), which became part of the pantheon of presolar grains. Other unusual isotopic ratios within these grains reveal many specific aspects of explosive nucleosynthesis.

4.4.4 Cosmic ray spallation

Main article: Cosmic ray spallation

Cosmic ray spallation process reduces the atomic weight of interstellar matter by the impact with cosmic rays, to produce some of the lightest elements present in the universe (though not a significant amount of deuterium). Most notably spallation is believed to be responsible for the generation of almost all of ^{3}He and the elements lithium, beryllium, and boron, although some 7Li and 7Be are thought to have been produced in the Big Bang. The spallation process results from the impact of cosmic rays (mostly fast protons) against the interstellar medium. These impacts fragment carbon, nitrogen, and oxygen nuclei present. The process results in the light elements beryllium, boron, and lithium in cosmos at much greater abundances than they are within solar atmospheres. The light elements ^{1}H and ^{4}He nuclei are not a product of spallation and are represented in the cosmos with approximately primordial abundance.

Beryllium and boron are not significantly produced by stellar fusion processes, due to the instability of any ^{8}Be formed from two ^{4}He nuclei.

4.5 Empirical evidence

Theories of nucleosynthesis are tested by calculating isotope abundances and comparing those results with observed results. Isotope abundances are typically calculated from the transition rates between isotopes in a network. Often these calculations can be simplified as a few key reactions control the rate of other reactions.

4.6 Minor mechanisms and processes

Very small amounts of certain nuclides are produced on Earth by artificial means. Those are our primary source, for example, of technetium. However, some nuclides are also produced by a number of natural means that have continued after primordial elements were in place. These often act to produce new elements in ways that can be used to date rocks or to trace the source of geological processes. Although these processes do not produce the nuclides in abundance, they are assumed to be the entire source of the existing natural supply of those nuclides.

These mechanisms include:

- Radioactive decay may lead to radiogenic daughter nuclides. The nuclear decay of many long-lived primordial isotopes, especially uranium-235, uranium-238, and thorium-232 produce many intermediate daughter nuclides, before they too finally decay to isotopes of lead. The Earth's natural supply of elements like radon and polonium is via this mechanism. The atmosphere's supply of argon-40 is due mostly to the radioactive decay of potassium-40 in the time since the formation of the Earth. Little of the atmospheric argon is primordial. Helium-4 is produced by alpha-decay, and the helium trapped in Earth's crust is also mostly non-primordial. In other types of radioactive decay, such as cluster decay, larger species of nuclei are ejected (for example, neon-20), and these eventually become newly formed stable atoms.

- Radioactive decay may lead to spontaneous fission. This is not cluster decay, as the fission products may be split among nearly any type of atom. Thorium-232, uranium-235, and uranium-238 are primordial isotopes that undergo spontaneous fission. Natural technetium and promethium are produced in this manner.

- Nuclear reactions. Naturally-occurring nuclear reactions powered by radioactive decay give rise to so-called nucleogenic nuclides. This process happens when an energetic particle from a radioactive decay, often an alpha particle, reacts with a nucleus of another atom to change the nucleus into another nuclide. This process may also cause the production of further subatomic particles, such as neutrons. Neutrons can also be produced in spontaneous fission and by neutron emission. These neutrons can then go on to produce other nuclides via neutron-induced fission, or by neutron capture. For example, some stable isotopes such as neon-21 and neon-22 are produced by several routes of nucleogenic synthesis, and thus only part of their abundance is primordial.

- Nuclear reactions due to cosmic rays. By convention, these reaction-products are not termed "nucleogenic" nuclides, but rather cosmogenic nuclides. Cosmic rays continue to produce new elements on Earth by the same cosmogenic processes discussed above that produce primordial beryllium and boron. One important example is carbon-14, produced from nitrogen-14 in the atmosphere by cosmic rays. Iodine-129 is another example.

In addition to artificial processes, it is postulated that neutron star collision is the main source of elements heavier than iron.[20]

4.7 See also

- Stellar evolution

- Supernova nucleosynthesis

- Cosmic dust

- Metallicity

4.8 References

[1] Donald D. Clayton, *Handbook of isotopes in the cosmos*, Cambridge University Press (Cambridge 2003)

[2] A.S. Eddington, The Internal Constitution of the Stars, *The Observatory*, **43**, 341 (1920) http://adsabs.harvard.edu/abs/1920Obs. ...43..341E

[3] A.S. Eddington, The Internal Constitution of the Stars, *Nature*, **106**, 106 (1920) http://adsabs.harvard.edu/abs/1920Natur.106. ..14E

[4] Actually, before the war ended, he learned abut the problem of spherical implosion of plutonium in the Manhattan project. He saw an analogy between the plutonium fission reaction and the newly discovered supernovae, and he was able to show that exploding super novae produced all of the elements in the same proportion as existed on Earth. He felt that he had accidentally fallen into a subject that would make his career. Autobiography William A. Fowler

[5] H.E. Suess and H.C. Urey, Abundances of the elements, *Revs. Mod. Phys.*, **28**, 53 (1957)

[6] Donald D. Clayton, *Handbook of isotopes in the cosmos*, Cambridge University Press (Cambridge U.K. 2003)

[7] Massimo S. Stiavelli. From First Light to Reionization. John Wiley & Sons, Apr 22, 2009. Pg 8.

[8] Donald D. Clayton, *Principles of Stellar Evolution and Nucleosynthesis*, McGraw-Hill (New York 1968) Chapter 5; reissued by University of Chicago Press (Chicago 1883)

[9] D.D. Clayton, W.A. Fowler, T. Hull and B. Zimmerman, Neutron capture chains in heavy element synthesis, *Ann. Phys.*, **12**, 331-408 (1961); Donald D. Clayton, *Principles of Stellar Evolution and Nucleosynthesis*, McGraw-Hill (New York 1968) Chapter 7

[10] S.Paul W.Merrill(1952). "Spectroscopic Observations of Stars of Class S".*The Astrophysical Journal***116**:21.Bibcode:19521M. doi:10.1086/145589.

[11] Donald D. Clayton and L. R. Nittler (2004). "Astrophysics with Presolar Stardust". *Annual Review of Astronomy and Astrophysics* **42** (1): 39–78. Bibcode:2004ARA&A..42...39C. doi:10.1146/annurev.astro.42.053102.134022.

[12] D. Bodansky, Donald D. Clayton, and W. A. Fowler, (1968) Nuclear quasi-equilibrium during silicon burning, *Astrophys. J. Suppl.* No. 148, **16**, 299-371

[13] See also Chapter 7 of Donald D. Clayton, *Principles of Stellar Evolution and Nucleosynthesis*, McGraw-Hill, New York (1968)

[14] Donald D. Clayton, Hoyle's Equation, *Science*, **318**, 1876-77 (2007)

[15] P.A.Seeger, W. A. Fowler, and Donald D. Clayton, Nucleosynthesis of heavy elements by neutron capture, *Astrophys. J. Suppl*, **11**, 121-66, (1965)

[16] D. Bodansky, Donald D. Clayton, and W. A. Fowler, (1968) Nuclear quasi-equilibrium during silicon burning, *Astrophys. J. Suppl.* No. 148, **16** 299-371

[17] Donald D. Clayton, Stirling A. Colgate and G. J. Fishman (1969) Gamma ray lines from young supernova remnants, *Astrophys. J..* **155** 175

[18] D. D. Clayton; S.A. Colgate; G.J. Fishman (1969). "Gamma ray lines from young supernova remnants". *The Astrophysical Journal* **155**: 75–82. Bibcode:1969ApJ...155...75C. doi:10.1086/149849.

[19] D. D. Clayton; L. R.Nittler (2004). "Astrophysics with Presolar stardust". *Annual Reviews of Astronomy and Astrophysics* **42** (1): 39–78. Bibcode:2004ARA&A..42...39C. doi:10.1146/annurev.astro.42.053102.134022.

[20] Stromberg, Joseph. "All the Gold in the Universe Could Come From the Collisions of Neutron Stars". *Smithsonian*. Retrieved 27 April 2014.

4.9   Further reading

- E. M. Burbidge, G. R. Burbidge, W. A. Fowler, F. Hoyle, *Synthesis of the Elements in Stars*, Rev. Mod. Phys. 29 (1957) 547 (article at the Physical Review Online Archive (subscription required)).

- M. Meneguzzi, J. Audouze, H. Reeves, « The production of the elements Li, Be, B by galactic cosmic rays in space and its relation with stellar observations », Astronomy and Astrophysics, vol. 15, 1971, p. 337–359

- F. Hoyle, Monthly Notices Roy. Astron. Soc. 106, 366 (1946)

- F. Hoyle, Astrophys. J. Suppl. 1, 121 (1954)

- D. D. Clayton, "Principles of Stellar Evolution and Nucleosynthesis", McGraw-Hill, 1968; University of Chicago Press, 1983, ISBN 0-226-10952-6

- C. E. Rolfs, W. S. Rodney, *Cauldrons in the Cosmos*, Univ. of Chicago Press, 1988, ISBN 0-226-72457-3.

- D. D. Clayton, "Handbook of Isotopes in the Cosmos", Cambridge University Press, 2003, ISBN 0-521-82381-1.

- C. Iliadis, "Nuclear Physics of Stars", Wiley-VCH, 2007, ISBN 978-3-527-40602-9

Chapter 5

Nuclear physics

For other uses, see Nuclear Physics (disambiguation).

Nuclear physics is the field of physics that studies the constituents and interactions of atomic nuclei. The most commonly known applications of nuclear physics are nuclear power generation but the research has provided application in many fields, including those in nuclear medicine and magnetic resonance imaging, nuclear weapons, ion implantation in materials engineering, and radiocarbon dating in geology and archaeology.

The field of particle physics evolved out of nuclear physics and is typically taught in close association with nuclear physics.

5.1   History

The history of nuclear physics as a discipline distinct from atomic physics starts with the discovery of radioactivity by Henri Becquerel in 1896,[1] while investigating phosphorescence in uranium salts.[2] The discovery of the electron by J. J. Thomson[3] a year later was an indication that the atom had internal structure. At the beginning of the 20th century the accepted model of the atom was J. J. Thomson's *plum pudding* model in which the atom was a large positively charged ball with small negatively charged electrons embedded inside of it.

In the years that followed, the phenomenon of radioactivity was extensively investigated, notably by the husband and wife team of Pierre Curie and Marie Curie and by Ernest Rutherford and his collaborators. By the turn of the century physicists had also discovered three types of radiation emanating from atoms, which they named alpha, beta, and gamma radiation. Experiments in 1911 by Otto Hahn, and by James Chadwick in 1914 discovered that the beta decay spectrum was continuous rather than discrete. That is, electrons were ejected from the atom with a range of energies, rather than the discrete amounts of energies that were observed in gamma and alpha decays. This was a problem for nuclear physics at the time, because it seemed to indicate that energy was not conserved in these decays.

The 1903 Nobel Prize in Physics was awarded jointly to Becquerel for his discovery and to Pierre Curie and Marie Curie for their subsequent research into radioactivity. Rutherford was awarded the Nobel Prize in Chemistry in 1908 for his 'investigations into the disintegration of the elements and the chemistry of radioactive substances'.

In 1905, Albert Einstein formulated the idea of mass–energy equivalence. While the work on radioactivity by Becquerel and Marie Curie predates this, an explanation of the source of the energy of radioactivity would have to wait for the discovery that the nucleus itself was composed of smaller constituents, the nucleons.

5.1.1   Rutherford's team discovers the nucleus

In 1907 Ernest Rutherford published "Radiation of the α Particle from Radium in passing through Matter."[4] Hans Geiger expanded on this work in a communication to the Royal Society[5] with experiments he and Rutherford had done, passing α particles through air, aluminum foil and gold leaf. More work was published in 1909 by Geiger and Marsden[6] and further greatly expanded work was published in 1910 by Geiger.[7] In 1911-1912 Rutherford went before the Royal Society to explain the experiments and propound the new theory of the atomic nucleus as we now understand it.

The key experiment behind this announcement happened in 1910 at the University of Manchester, as Ernest Rutherford's team performed a remarkable experiment in which Hans Geiger and Ernest Marsden under his supervision fired alpha particles (helium nuclei) at a thin film of gold foil. The plum pudding model predicted that the alpha particles should come out of the foil with their trajectories being at most slightly bent. Rutherford had the idea to instruct his team to look for something that shocked him to actually observe: a few particles were scattered through large angles, even completely backwards, in some cases. He likened it to firing a bullet at tissue paper and having it bounce off. The discovery, beginning with Rutherford's analysis of the data in 1911, eventually led to the Rutherford model of the atom, in which the atom has a very small, very dense nucleus containing most of its mass, and consisting of heavy positively charged particles with embedded electrons in order to balance out the charge (since the neutron was unknown). As an example, in this model (which is not the modern one) nitrogen-14 consisted of a nucleus with 14 protons and 7 electrons (21 total particles), and the nucleus was surrounded by 7 more orbiting electrons.

The Rutherford model worked quite well until studies of nuclear spin were carried out by Franco Rasetti at the California Institute of Technology in 1929. By 1925 it was known that protons and electrons had a spin of 1/2, and in the Rutherford model of nitrogen-14, 20 of the total 21 nuclear particles should have paired up to cancel each other's spin, and the final odd particle should have left the nucleus with a net spin of 1/2. Rasetti discovered, however, that nitrogen-14 had a spin of 1.

5.1.2   James Chadwick discovers the neutron

In 1932 Chadwick realized that radiation that had been observed by Walther Bothe, Herbert Becker, Irène and Frédéric Joliot-Curie was actually due to a neutral particle of about the same mass as the proton, that he called the neutron (following a suggestion about the need for such a particle, by Rutherford).[8] In the same year Dmitri Ivanenko suggested that neutrons were in fact spin 1/2 particles and that the nucleus contained neutrons to explain the mass not due to protons, and that there were no electrons in the nucleus — only protons and neutrons. The neutron spin immediately solved the problem of the spin of nitrogen-14, as the one unpaired proton and one unpaired neutron in this model, each contribute a spin of 1/2 in the same direction, for a final total spin of 1.

With the discovery of the neutron, scientists at last could calculate what fraction of binding energy each nucleus had, from comparing the nuclear mass with that of the protons and neutrons which composed it. Differences between nuclear masses were calculated in this way and—when nuclear reactions were measured—were found to agree with Einstein's calculation of the equivalence of mass and energy to high accuracy (within 1 percent as of in 1934).

5.1.3   Proca's equations of the massive vector boson field

Alexandru Proca was the first to develop and report the massive vector boson field equations and a theory of the mesonic field of nuclear forces. Proca's equations were known to Wolfgang Pauli[9] who mentioned the equations in his Nobel address, and they were also known to Yukawa, Wentzel, Taketani, Sakata, Kemmer, Heitler, and Fröhlich who appreciated the content of Proca's equations for developing a theory of the atomic nuclei in Nuclear Physics.[10][11][12][13][14]

5.1.4   Yukawa's meson postulated to bind nuclei

In 1935 Hideki Yukawa [15] proposed the first significant theory of the strong force to explain how the nucleus holds together. In the Yukawa interaction a virtual particle, later called a meson, mediated a force between all nucleons, including protons and neutrons. This force explained why nuclei did not disintegrate under the influence of proton repulsion, and it also gave an explanation of why the attractive strong force had a more limited range than the electromagnetic repulsion between protons. Later, the discovery of the pi meson showed it to have the properties of Yukawa's particle.

With Yukawa's papers, the modern model of the atom was complete. The center of the atom contains a tight ball of neutrons and protons, which is held together by the strong nuclear force, unless it is too large. Unstable nuclei may undergo alpha decay, in which they emit an energetic helium nucleus, or beta decay, in which they eject an electron (or positron). After one of these decays the resultant nucleus may be left in an excited state, and in this case it decays to its ground state by emitting high energy photons (gamma decay).

The study of the strong and weak nuclear forces (the latter explained by Enrico Fermi via Fermi's interaction in 1934) led physicists to collide nuclei and electrons at ever higher energies. This research became the science of particle physics, the crown jewel of which is the standard model of particle physics which describes the strong, weak, and electromagnetic forces.

5.2 Modern nuclear physics

Main articles: Liquid-drop model, Nuclear shell model and Nuclear structure

A heavy nucleus can contain hundreds of nucleons which means that with some approximation it can be treated as a classical system, rather than a quantum-mechanical one. In the resulting liquid-drop model,[16] the nucleus has an energy which arises partly from surface tension and partly from electrical repulsion of the protons. The liquid-drop model is able to reproduce many features of nuclei, including the general trend of binding energy with respect to mass number, as well as the phenomenon of nuclear fission.

Superimposed on this classical picture, however, are quantum-mechanical effects, which can be described using the nuclear shell model, developed in large part by Maria Goeppert Mayer[17] and J. Hans D. Jensen.[18] Nuclei with certain numbers of neutrons and protons (the magic numbers 2, 8, 20, 28, 50, 82, 126, ...) are particularly stable, because their shells are filled.

Other more complicated models for the nucleus have also been proposed, such as the interacting boson model, in which pairs of neutrons and protons interact as bosons, analogously to Cooper pairs of electrons.

Much of current research in nuclear physics relates to the study of nuclei under extreme conditions such as high spin and excitation energy. Nuclei may also have extreme shapes (similar to that of Rugby balls or even pears) or extreme neutron-to-proton ratios. Experimenters can create such nuclei using artificially induced fusion or nucleon transfer reactions, employing ion beams from an accelerator. Beams with even higher energies can be used to create nuclei at very high temperatures, and there are signs that these experiments have produced a phase transition from normal nuclear matter to a new state, the quark–gluon plasma, in which the quarks mingle with one another, rather than being segregated in triplets as they are in neutrons and protons.

5.2.1 Nuclear decay

Main article: Radioactivity

Eighty elements have at least one stable isotope never observed to decay, amounting to a total of about 254 stable isotopes. However, thousands of isotopes have been characterized as unstable. These radioisotopes decay over time scales ranging from fractions of a second to weeks, years, billions of years, or even trillions of years.

The stability of a nucleus is highest when it falls into a certain range or balance of composition of neutrons and protons; too few or too many neutrons may cause it to decay. For example, in beta decay a nitrogen−16 atom (7 protons, 9 neutrons) is converted to an oxygen−16 atom (8 protons, 8 neutrons) within a few seconds of being created. In this decay a neutron in the nitrogen nucleus is converted into a proton and an electron and an antineutrino by the weak interaction. The element is transmuted to another element by acquiring the created proton.

In alpha decay (which typically occurs in the heaviest nuclei) the radioactive element decays by emitting a helium nucleus (2 protons and 2 neutrons), giving another element, plus helium-4. In many cases this process continues through several steps of this kind, including other types of decays, until a stable element is formed.

In gamma decay, a nucleus decays from an excited state into a lower energy state, by emitting a gamma ray. The element is not changed to another element in the process (no nuclear transmutation is involved).

Other more exotic decays are possible (see the main article). For example, in internal conversion decay, the energy from an excited nucleus may be used to eject one of the inner orbital electrons from the atom, in a process which produces high speed electrons, but is not beta decay, and (unlike beta decay) does not transmute one element to another.

5.2.2 Nuclear fusion

In nuclear fusion, two low mass nuclei come into very close contact with each other, so that the strong force fuses them. It requires a large amount of energy to overcome the repulsion between the nuclei for the strong or nuclear forces to produce this effect, therefore nuclear fusion can only take place at very high temperatures or high pressures. Once the process succeeds, a very large amount of energy is released and the combined nucleus assumes a lower energy level. The binding energy per nucleon increases with mass number up until nickel−62. Stars like the Sun are powered by the fusion of four protons into a helium nucleus, two positrons, and two neutrinos. The uncontrolled fusion of hydrogen into helium is known as thermonuclear runaway. A frontier in current research at various institutions, for example the Joint European Torus (JET) and ITER, is the development of an economically viable method of using energy from a *controlled* fusion reaction. Natural nuclear fusion is the origin of the light and energy produced by the core of all stars including our own sun.

5.2.3 Nuclear fission

Nuclear fission is the reverse process of fusion. For nuclei heavier than nickel-62 the binding energy per nucleon decreases with the mass number. It is therefore possible for energy to be released if a heavy nucleus breaks apart into two lighter ones.

The process of alpha decay is in essence a special type of spontaneous nuclear fission. This process produces a highly asymmetrical fission because the four particles which make up the alpha particle are especially tightly bound to each other, making production of this nucleus in fission particularly likely.

For certain of the heaviest nuclei which produce neutrons on fission, and which also easily absorb neutrons to initiate fission, a self-igniting type of neutron-initiated fission can be obtained, in a so-called chain reaction. Chain reactions were known in chemistry before physics, and in fact many familiar processes like fires and chemical explosions are chemical chain reactions. The fission or "nuclear" chain-reaction, using fission-produced neutrons, is the source of energy for nuclear power plants and fission type nuclear bombs, such as those detonated by the United States in Hiroshima and Nagasaki, Japan, at the end of World War II. Heavy nuclei such as uranium and thorium may also undergo spontaneous fission, but they are much more likely to undergo decay by alpha decay.

For a neutron-initiated chain-reaction to occur, there must be a critical mass of the element present in a certain space under certain conditions. The conditions for the smallest critical mass require the conservation of the emitted neutrons and also their slowing or moderation so there is a greater cross-section or probability of them initiating another fission. In two regions of Oklo, Gabon, Africa, natural nuclear fission reactors were active over 1.5 billion years ago.[19] Measurements of natural neutrino emission have demonstrated that around half of the heat emanating from the Earth's core results from radioactive decay. However, it is not known if any of this results from fission chain-reactions.

5.2.4 Production of "heavy" elements (atomic number greater than five)

Main article: nucleosynthesis

According to the theory, as the Universe cooled after the big bang it eventually became possible for common subatomic particles as we know them (neutrons, protons and electrons) to exist. The most common particles created in the big bang which are still easily observable to us today were protons and electrons (in equal numbers). The protons would eventually form hydrogen atoms. Almost all the neutrons created in the Big Bang were absorbed into helium-4 in the first three minutes after the Big Bang, and this helium accounts for most of the helium in the universe today (see Big Bang nucleosynthesis).

Some fraction of elements beyond helium were created in the Big Bang, as the protons and neutrons collided with each other (lithium, beryllium, and perhaps some boron), but all of the "heavier elements" (carbon, element number 6, and elements of greater atomic number) that we see today, were created inside of stars during a series of fusion stages, such as the proton-proton chain, the CNO cycle and the triple-alpha process. Progressively heavier elements are created during the evolution of a star.

Since the binding energy per nucleon peaks around iron, energy is only released in fusion processes occurring below this point. Since the creation of heavier nuclei by fusion costs energy, nature resorts to the process of neutron capture. Neutrons (due to their lack of charge) are readily absorbed by a nucleus. The heavy elements are created by either a *slow* neutron capture process (the so-called *s* process) or by the *rapid*, or *r* process. The *s* process occurs in thermally pulsing stars (called AGB, or asymptotic giant branch stars) and takes hundreds to thousands of years to reach the heaviest elements of lead and bismuth. The *r* process is thought to occur in supernova explosions because the conditions of high temperature, high neutron flux and ejected matter are present. These stellar conditions make the successive neutron captures very fast, involving very neutron-rich species which then beta-decay to heavier elements, especially at the so-called waiting points that correspond to more stable nuclides with closed neutron shells (magic numbers).

5.3   See also

- Isomeric shift
- Neutron-degenerate matter
- Nuclear matter
- Nuclear model
- Nuclear reactor physics
- QCD matter

5.4   References

[1] B. R. Martin (2006). *Nuclear and Particle Physics*. John Wiley & Sons, Ltd. ISBN 0-470-01999-9.

[2] Henri Becquerel (1896). "Sur les radiations émises par phosphorescence". *Comptes Rendus* **122**: 420–421.

[3] J.J. Thomson (1897) 'The Electrician *39, 104*

[4] *Philosophical Magazine* (**12**, p 134-46)

[5] *Proc. Roy. Soc.* July 17, 1908

[6] *Proc. Roy. Soc.* **A82**: 495–500. Missing or empty |title= (help)

[7] H. Geiger, Roy. Soc. Proc. vol. LXXXIII (1910) 492

[8] J. Chadwick, Nature 192 (1932) 312

[9] W. Pauli, *Nobel lecture*, December 13, 1946.

[10] Poenaru, Dorin N.; Calboreanu, Alexandru. "Alexandru Proca (1897-1955) and his equation of the massive vector boson field". *Europhysics News* **37** (5): 25–27. Bibcode:2006ENews..37...24P. doi:10.1051/epn:2006504.

[11] *G. A. Proca, Alexandre Proca. Oeuvre Scientifique Publiée*, S.I.A.G., Rome, 1988.

[12] Vuille, C.; Ipser, J.; Gallagher, J. (2002). "Einstein-Proca model, micro black holes, and naked singularities". *General Relativity and Gravitation* **34**: 689.

[13] Scipioni, R. (1999). "Isomorphism between non-Riemannian gravity and Einstein-Proca-Weyl theories extended to a class of scalar gravity theories". *Class. Quantum Gravity* **16**: 2471–2478. arXiv:gr-qc/9905022. Bibcode:1999CQGra..16.2471S. doi:10.1088/0264-9381/16/7/320.

[14] Tucker, R. W.; Wang, C. (1997). "An Einstein-Proca-fluid model for dark matter gravitational interactions", *Nucl. Phys. B -". Proc. suppl* **57**: 259–262. Bibcode:1997NuPhS..57..259T. doi:10.1016/s0920-5632(97)00399-x.

[15] On the Interaction of Elementary Particles I. Proceedings of the Physico-Mathematical Society of Japan. 3rd Series Vol. 17 (1935) p. 48-57

[16] J.M.Blatt and V.F.Weisskopf, Theoretical Nuclear Physics, Springer, 1979, VII.5

[17] M.G. Mayer, Physical Review 75 (1949) 1969

[18] O. Haxel, J.H.D. Jensen, H.E. Suess, Physical Review, 75 (1949) 1766

[19] Meshik, A. P. (November 2005). "The Workings of an Ancient Nuclear Reactor". *Scientific American*. Retrieved 2014-01-04.

5.5 Bibliography

- Nuclear Physics by Irving Kaplan 2nd edition1962 Addison-Wesley

- General Chemistry by Linus Pauling 1970 Dover Pub. ISBN 0-486-65622-5

- Introductory Nuclear Physics by Kenneth S. Krane Pub. Wiley

- N.D. Cook (2010). *Models of the Atomic Nucleus* (2nd ed.). Springer. pp. xvi & 324. ISBN 978-3-642-14736-4.

- Ahmad, D.Sc., Ishfaq; American Institute of Physics (1996). *Physics of particles and nuclei*. 1-3 **27** (3rd ed.). University of California: American Institute of Physics Press. p. 124.

5.6 External links

- Ernest Rutherford's biography at the American Institute of Physics

- American Physical Society Division of Nuclear Physics

- Amcrican Nuclcar Socicty

- Boiling Water Reactor Plant, BWR Simulator Program

- Annotated bibliography on nuclear physics from the Alsos Digital Library for Nuclear Issues

- Nucleonica ..web driven nuclear science

- Nuclear science wiki

- Nuclear Data Services - IAEA

Chapter 6

Rutherford model

The **Rutherford model** is a model of the atom devised by Ernest Rutherford. Rutherford directed the famous Geiger–Marsden experiment in 1909 which suggested, upon Rutherford's 1911 analysis, that the so-called "plum pudding model" of J. J. Thomson of the atom was incorrect. Rutherford's new model[1] for the atom, based on the experimental results, contained the new features of a relatively high central charge concentrated into a very small volume in comparison to the rest of the atom and with this central volume also containing the bulk of the atomic mass of the atom. This region would be named the "nucleus" of the atom in later years.

6.1   Experimental basis for the model

Rutherford overturned Thomson's model in 1911 with his well-known gold foil experiment in which he demonstrated that the atom has a tiny, heavy nucleus. Rutherford designed an experiment to use the alpha particles emitted by a radioactive element as probes to the unseen world of atomic structure.

Rutherford presented his own physical model for subatomic structure, as an interpretation for the unexpected experimental results. In it, the atom is made up of a central charge (this is the modern atomic nucleus, though Rutherford did not use the term "nucleus" in his paper) surrounded by a cloud of (presumably) orbiting electrons. In this May 1911 paper, Rutherford only commits himself to a small central region of very high positive or negative charge in the atom.

> For concreteness, consider the passage of a high speed α particle through an atom having a positive central charge N e, and surrounded by a compensating charge of N electrons.[2]

From purely energetic considerations of how far particles of known speed would be able to penetrate toward a central charge of 100 e, Rutherford was able to calculate that the radius of his gold central charge would need to be less (how much less could not be told) than 3.4×10^{-14} meters. This was in a gold atom known to be 10^{-10} meters or so in radius—a very surprising finding, as it implied a strong central charge less than 1/3000th of the diameter of the atom.

The Rutherford model served to concentrate a great deal of the atom's charge and mass to a very small core, but didn't attribute any structure to the remaining electrons and remaining atomic mass. It did mention the atomic model of Hantaro Nagaoka, in which the electrons are arranged in one or more rings, with the specific metaphorical structure of the stable rings of Saturn. The plum pudding model of J.J. Thomson also had rings of orbiting electrons. Jean Baptiste Perrin claimed in his Nobel Lecture [3] that he was the first one to suggest the model in his paper dated 1901.

The Rutherford paper suggested that the central charge of an atom might be "proportional" to its atomic mass in hydrogen mass units u (roughly 1/2 of it, in Rutherford's model). For gold, this mass number is 197 (not then known to great accuracy) and was therefore modeled by Rutherford to be possibly 196 u. However, Rutherford did not attempt to make the direct connection of central charge to atomic number, since gold's "atomic number" (at *that* time merely its place number in the periodic table) was 79, and Rutherford had modeled the charge to be about + 100 units (he had actually suggested 98 units of positive charge, to make half of 196). Thus, Rutherford did not formally suggest the two numbers (periodic table place, 79, and nuclear charge, 98 or 100) might be exactly the same.

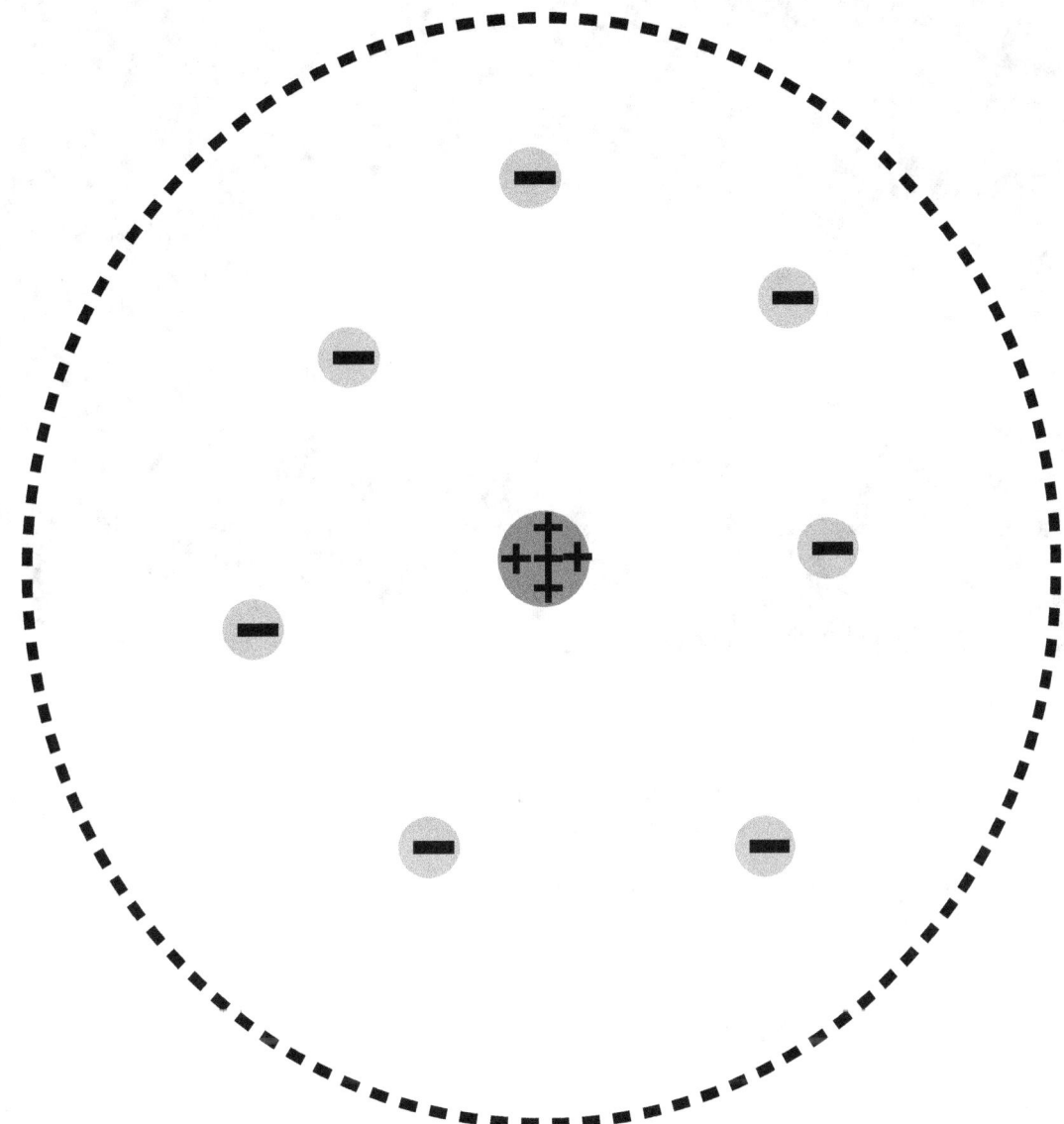

Basic diagram of the atomic planetary model: electrons in green and nucleus in red

A month after Rutherford's paper appeared, the proposal regarding the exact identity of atomic number and nuclear charge *was* made by Antonius van den Broek, and later confirmed experimentally within two years, by Henry Moseley.

6.2   Key points

- The atom's electron cloud does not influence alpha particle scattering.

- Much of an atom's positive charge is concentrated in a relatively tiny volume at the center of the atom, known today as the nucleus. The magnitude of this charge is proportional to (up to a charge number that can be approximately half of) the atom's atomic mass - the remaining mass is now known to be mostly attributed to neutrons. This concentrated central mass and charge is responsible for deflecting both alpha and beta particles.

- The mass of heavy atoms such as gold is mostly concentrated in the central charge region, since calculations show

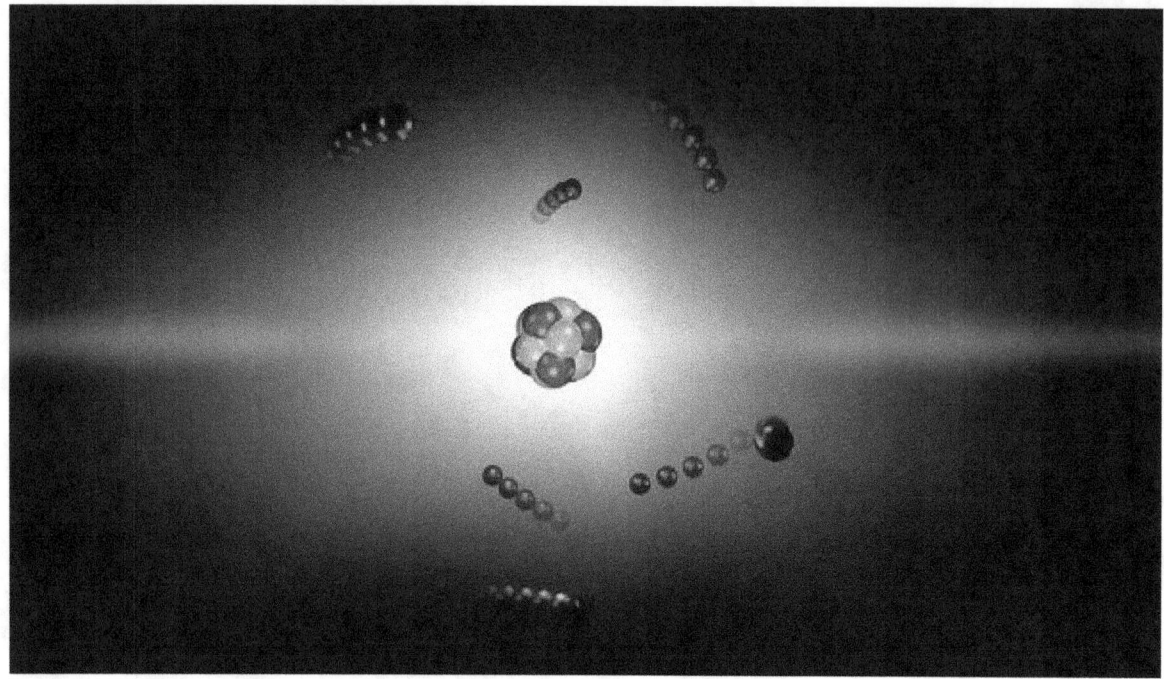

3D animation of an atom incorporating the Rutherford model

it is not deflected or moved by the high speed alpha particles, which have very high momentum in comparison to electrons, but not with regard to a heavy atom as a whole.

- The atom itself is about 100,000 (10^5) times the diameter of the nucleus.[4] This could be related to putting an apple in the middle of a football field.[5]

6.3 Contribution to modern science

After Rutherford's discovery, scientists started to realize that the atom is not ultimately a single particle, but is made up of far smaller subatomic particles. Subsequent research determined the exact atomic structure which led to Rutherford's gold foil experiment. Scientists eventually discovered that atoms have a positively charged nucleus (with an exact atomic number of charges) in the center, with a radius of about 1.2 x 10^{-15} meters x [Atomic Mass Number]$^{1/3}$. Electrons were found to be even smaller.

Later, scientists found the expected number of electrons (the same as the atomic number) in an atom by using X-rays. When an X-ray passes through an atom, some of it is scattered while the rest passes through the atom. Since the X-ray loses its intensity primarily due to scattering at electrons, by noting the rate of decrease in X-ray intensity, the number of electrons contained in an atom can be accurately estimated.

6.4 Symbolism

See also Bohr model, which applies just as well to the section below.

Rutherford's model deferred to the idea of many electrons in rings, per Nagaoka. However, once Niels Bohr modified this view into a picture of just a few planet-like electrons for light atoms, the Rutherford-Bohr model caught the imagination of the public. It has since continually been used as a symbol for atoms and even for "atomic" energy (even though this is more properly considered nuclear energy). Examples of its use over the past century include:

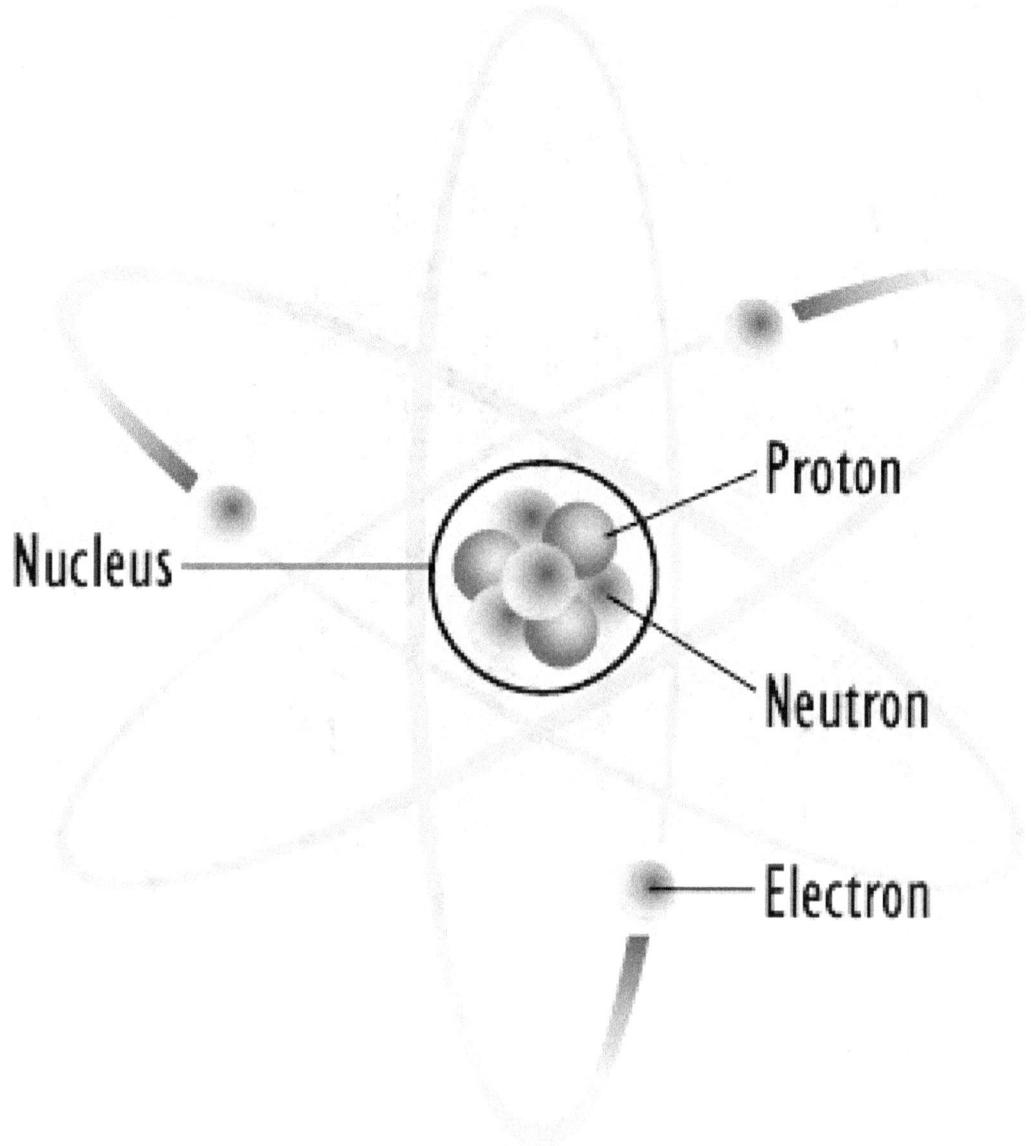

Generic atomic planetary model.

- The logo of the United States Atomic Energy Commission, which was in part responsible for its later usage in relation to nuclear fission technology in particular.

- The flag of the International Atomic Energy Agency is a Rutherford atom, enclosed in olive branches.

- The US minor league baseball Albuquerque Isotopes' logo is a Rutherford atom, with the electron orbits forming an A.

- A similar symbol, the Atomic whirl, was chosen as the symbol for the American Atheists, and has come to be used as a symbol of atheism in general.

- The Unicode Miscellaneous Symbols codepoint U+269B (⚛) uses a Rutherford atom.

- On maps, it is generally used to indicate a nuclear power installation.

Shield of the U.S. Atomic Energy Commission

6.5 References

[1] Akhlesh Lakhtakia (Ed.); Salpeter, Edwin E. (1996). "Models and Modelers of Hydrogen". *American Journal of Physics* (World Scientific) **65** (9): 933. Bibcode:1997AmJPh..65..933L. doi:10.1119/1.18691. ISBN 981-02-2302-1.

[2] E. Rutherford, *The Scattering of α and β Particles by Matter and the Structure of the Atom*, Philosophical Magazine. Series 6, vol. **21**. May 1911

[3] 1926 Lecture for Nobel Prize in Physics

[4] Nicholas Giordano (1 January 2012). *College Physics: Reasoning and Relationships*. Cengage Learning. pp. 1051–. ISBN 1-285-22534-1.

[5] Constan, Zach (2010). "Learning Nuclear Science with Marbles". *The Physics Teacher* **48** (2): 114. doi:10.1119/1.3293660. ISSN 0031-921X.

6.6 External links

- Rutherford's Model by Raymond College
- Rutherford's Model by Kyushu University

Chapter 7

Standard Model

This article is about the Standard Model of particle physics. For other uses, see Standard model (disambiguation).
This article is a non-mathematical general overview of the Standard Model. For a mathematical description, see the article Standard Model (mathematical formulation).
For the Standard Model of Big Bang cosmology, Lambda-CDM model.

The **Standard Model** of particle physics is a theory concerning the electromagnetic, weak, and strong nuclear inter-

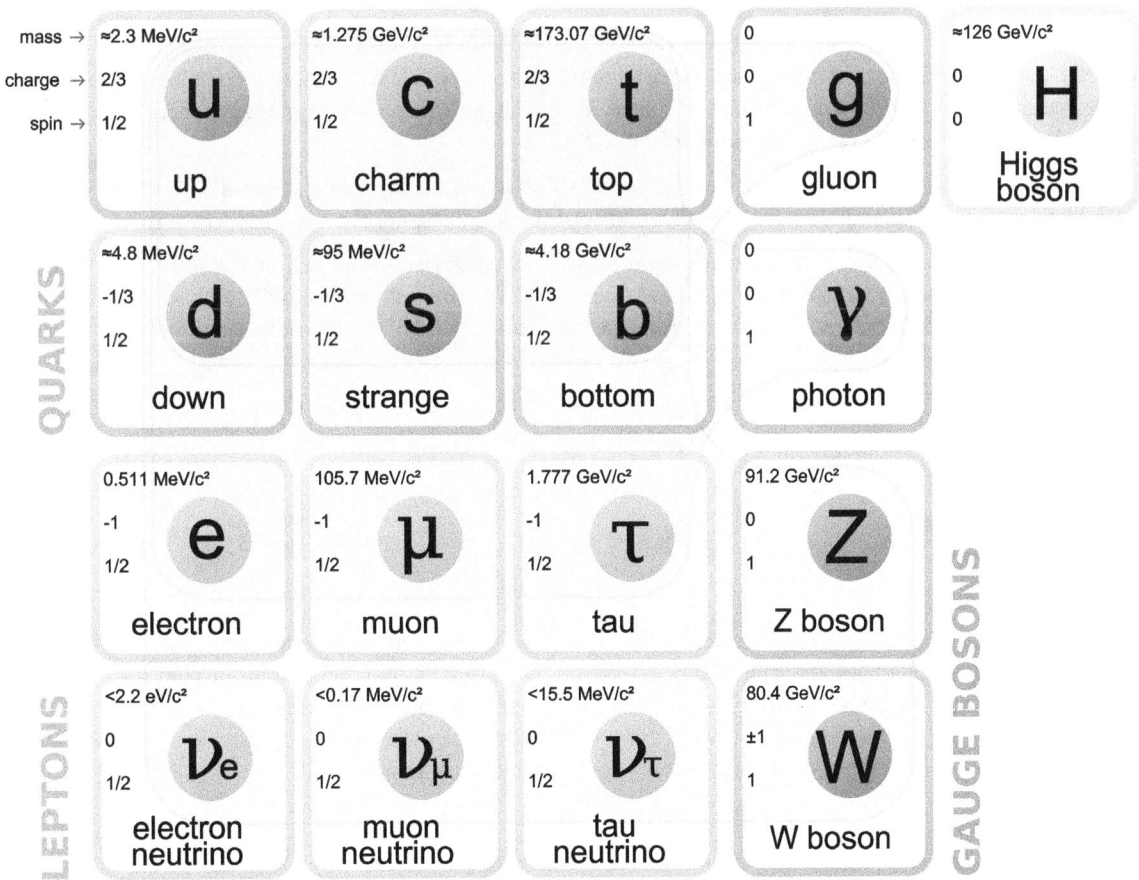

The Standard Model of elementary particles (more schematic depiction), with the three generations of matter, gauge bosons in the fourth column, and the Higgs boson in the fifth.

actions, as well as classifying all the subatomic particles known. It was developed throughout the latter half of the 20th century, as a collaborative effort of scientists around the world.[1] The current formulation was finalized in the mid-1970s upon experimental confirmation of the existence of quarks. Since then, discoveries of the top quark (1995), the tau neutrino (2000), and more recently the Higgs boson (2013), have given further credence to the Standard Model. Because of its success in explaining a wide variety of experimental results, the Standard Model is sometimes regarded as a "theory of almost everything".

Although the Standard Model is believed to be theoretically self-consistent[2] and has demonstrated huge and continued successes in providing experimental predictions, it does leave some phenomena unexplained and it falls short of being a complete theory of fundamental interactions. It does not incorporate the full theory of gravitation[3] as described by general relativity, or account for the accelerating expansion of the universe (as possibly described by dark energy). The model does not contain any viable dark matter particle that possesses all of the required properties deduced from observational cosmology. It also does not incorporate neutrino oscillations (and their non-zero masses).

The development of the Standard Model was driven by theoretical and experimental particle physicists alike. For theorists, the Standard Model is a paradigm of a quantum field theory, which exhibits a wide range of physics including spontaneous symmetry breaking, anomalies, non-perturbative behavior, etc. It is used as a basis for building more exotic models that incorporate hypothetical particles, extra dimensions, and elaborate symmetries (such as supersymmetry) in an attempt to explain experimental results at variance with the Standard Model, such as the existence of dark matter and neutrino oscillations.

7.1 Historical background

The first step towards the Standard Model was Sheldon Glashow's discovery in 1961 of a way to combine the electromagnetic and weak interactions.[4] In 1967 Steven Weinberg[5] and Abdus Salam[6] incorporated the Higgs mechanism[7][8][9] into Glashow's electroweak theory, giving it its modern form.

The Higgs mechanism is believed to give rise to the masses of all the elementary particles in the Standard Model. This includes the masses of the W and Z bosons, and the masses of the fermions, i.e. the quarks and leptons.

After the neutral weak currents caused by Z boson exchange were discovered at CERN in 1973,[10][11][12][13] the electroweak theory became widely accepted and Glashow, Salam, and Weinberg shared the 1979 Nobel Prize in Physics for discovering it. The W and Z bosons were discovered experimentally in 1981, and their masses were found to be as the Standard Model predicted.

The theory of the strong interaction, to which many contributed, acquired its modern form around 1973–74, when experiments confirmed that the hadrons were composed of fractionally charged quarks.

7.2 Overview

At present, matter and energy are best understood in terms of the kinematics and interactions of elementary particles. To date, physics has reduced the laws governing the behavior and interaction of all known forms of matter and energy to a small set of fundamental laws and theories. A major goal of physics is to find the "common ground" that would unite all of these theories into one integrated theory of everything, of which all the other known laws would be special cases, and from which the behavior of all matter and energy could be derived (at least in principle).[14]

7.3 Particle content

The Standard Model includes members of several classes of elementary particles (fermions, gauge bosons, and the Higgs boson), which in turn can be distinguished by other characteristics, such as color charge.

7.3.1 Fermions

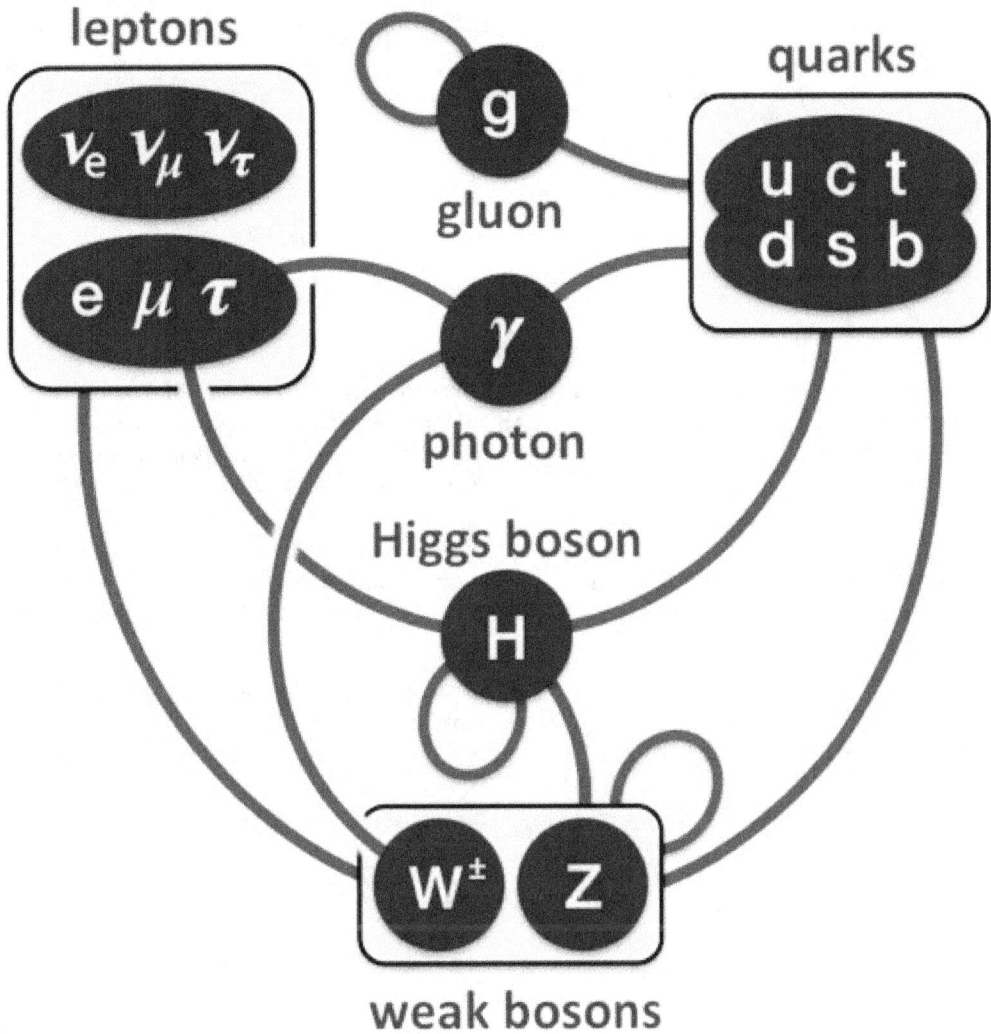

Summary of interactions between particles described by the Standard Model.

The Standard Model includes 12 elementary particles of spin-½ known as fermions. According to the spin-statistics theorem, fermions respect the Pauli exclusion principle. Each fermion has a corresponding antiparticle.

The fermions of the Standard Model are classified according to how they interact (or equivalently, by what charges they carry). There are six quarks (up, down, charm, strange, top, bottom), and six leptons (electron, electron neutrino, muon, muon neutrino, tau, tau neutrino). Pairs from each classification are grouped together to form a generation, with corresponding particles exhibiting similar physical behavior (see table).

The defining property of the quarks is that they carry color charge, and hence, interact via the strong interaction. A phenomenon called color confinement results in quarks being very strongly bound to one another, forming color-neutral composite particles (hadrons) containing either a quark and an antiquark (mesons) or three quarks (baryons). The familiar proton and the neutron are the two baryons having the smallest mass. Quarks also carry electric charge and weak isospin. Hence they interact with other fermions both electromagnetically and via the weak interaction.

The remaining six fermions do not carry colour charge and are called leptons. The three neutrinos do not carry electric charge either, so their motion is directly influenced only by the weak nuclear force, which makes them notoriously difficult to detect. However, by virtue of carrying an electric charge, the electron, muon, and tau all interact electromagnetically.

Each member of a generation has greater mass than the corresponding particles of lower generations. The first generation charged particles do not decay; hence all ordinary (baryonic) matter is made of such particles. Specifically, all atoms consist of electrons orbiting around atomic nuclei, ultimately constituted of up and down quarks. Second and third generations charged particles, on the other hand, decay with very short half lives, and are observed only in very high-energy environments. Neutrinos of all generations also do not decay, and pervade the universe, but rarely interact with baryonic matter.

7.3.2   Gauge bosons

In the Standard Model, gauge bosons are defined as force carriers that mediate the strong, weak, and electromagnetic fundamental interactions.

Interactions in physics are the ways that particles influence other particles. At a macroscopic level, electromagnetism allows particles to interact with one another via electric and magnetic fields, and gravitation allows particles with mass to attract one another in accordance with Einstein's theory of general relativity. The Standard Model explains such forces as resulting from matter particles exchanging other particles, generally referred to as *force mediating particles*. When a force-mediating particle is exchanged, at a macroscopic level the effect is equivalent to a force influencing both of them, and the particle is therefore said to have *mediated* (i.e., been the agent of) that force. The Feynman diagram calculations, which are a graphical representation of the perturbation theory approximation, invoke "force mediating particles", and when applied to analyze high-energy scattering experiments are in reasonable agreement with the data. However, perturbation theory (and with it the concept of a "force-mediating particle") fails in other situations. These include low-energy quantum chromodynamics, bound states, and solitons.

The gauge bosons of the Standard Model all have spin (as do matter particles). The value of the spin is 1, making them bosons. As a result, they do not follow the Pauli exclusion principle that constrains fermions: thus bosons (e.g. photons) do not have a theoretical limit on their spatial density (number per volume). The different types of gauge bosons are described below.

- Photons mediate the electromagnetic force between electrically charged particles. The photon is massless and is well-described by the theory of quantum electrodynamics.

- The W+, W−, and Z gauge bosons mediate the weak interactions between particles of different flavors (all quarks and leptons). They are massive, with the Z being more massive than the W±. The weak interactions involving the W± exclusively act on *left-handed* particles and *right-handed* antiparticles. Furthermore, the W± carries an electric charge of +1 and −1 and couples to the electromagnetic interaction. The electrically neutral Z boson interacts with both left-handed particles and antiparticles. These three gauge bosons along with the photons are grouped together, as collectively mediating the electroweak interaction.

- The eight gluons mediate the strong interactions between color charged particles (the quarks). Gluons are massless. The eightfold multiplicity of gluons is labeled by a combination of color and anticolor charge (e.g. red–antigreen).[nb 1] Because the gluons have an effective color charge, they can also interact among themselves. The gluons and their interactions are described by the theory of quantum chromodynamics.

The interactions between all the particles described by the Standard Model are summarized by the diagrams on the right of this section.

7.3.3   Higgs boson

Main article: Higgs boson

Standard Model Interactions
(Forces Mediated by Gauge Bosons)

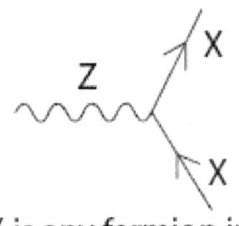

X is any fermion in
the Standard Model.

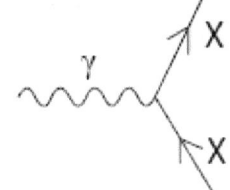

X is electrically charged.

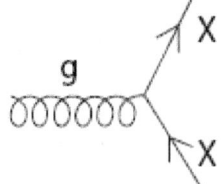

X is any quark.

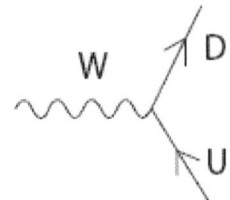

U is a up-type quark;
D is a down-type quark.

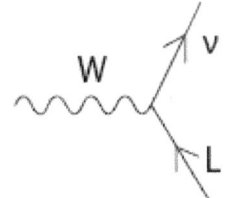

L is a lepton and ν is the
corresponding neutrino.

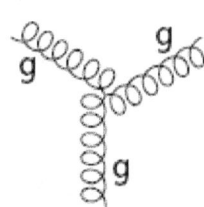

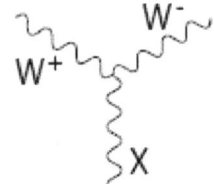

X is a photon or Z-boson.

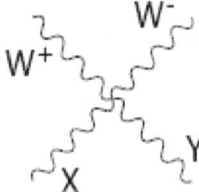

X and Y are any two
electroweak bosons such
that charge is conserved.

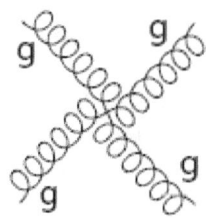

*The above interactions form the basis of the standard model. Feynman diagrams in the standard model are built from these vertices.
Modifications involving Higgs boson interactions and neutrino oscillations are omitted. The charge of the W bosons is dictated by the
fermions they interact with; the conjugate of each listed vertex (i.e. reversing the direction of arrows) is also allowed.*

The Higgs particle is a massive scalar elementary particle theorized by Robert Brout, François Englert, Peter Higgs, Gerald Guralnik, C. R. Hagen, and Tom Kibble in 1964 (see 1964 PRL symmetry breaking papers) and is a key building block in the Standard Model.[7][8][9][15] It has no intrinsic spin, and for that reason is classified as a boson (like the gauge bosons, which have integer spin).

The Higgs boson plays a unique role in the Standard Model, by explaining why the other elementary particles, except the photon and gluon, are massive. In particular, the Higgs boson explains why the photon has no mass, while the W and Z bosons are very heavy. Elementary particle masses, and the differences between electromagnetism (mediated by the photon) and the weak force (mediated by the W and Z bosons), are critical to many aspects of the structure of microscopic (and hence macroscopic) matter. In electroweak theory, the Higgs boson generates the masses of the leptons (electron, muon, and tau) and quarks. As the Higgs boson is massive, it must interact with itself.

Because the Higgs boson is a very massive particle and also decays almost immediately when created, only a very high-energy particle accelerator can observe and record it. Experiments to confirm and determine the nature of the Higgs boson using the Large Hadron Collider (LHC) at CERN began in early 2010, and were performed at Fermilab's Tevatron until its closure in late 2011. Mathematical consistency of the Standard Model requires that any mechanism capable of generating the masses of elementary particles become visible at energies above 1.4 TeV;[16] therefore, the LHC (designed to collide two 7 to 8 TeV proton beams) was built to answer the question of whether the Higgs boson actually exists.[17]

On 4 July 2012, the two main experiments at the LHC (ATLAS and CMS) both reported independently that they found a new particle with a mass of about 125 GeV/c^2 (about 133 proton masses, on the order of 10^{-25} kg), which is "consistent with the Higgs boson." Although it has several properties similar to the predicted "simplest" Higgs,[18] they acknowledged that further work would be needed to conclude that it is indeed the Higgs boson, and exactly which version of the Standard Model Higgs is best supported if confirmed.[19][20][21][22][23]

On 14 March 2013 the Higgs Boson was tentatively confirmed to exist.[24]

7.3.4 Total particle count

Counting particles by a rule that distinguishes between particles and their corresponding antiparticles, and among the many color states of quarks and gluons, gives a total of 61 elementary particles.[25]

7.4   Theoretical aspects

Main article: Standard Model (mathematical formulation)

7.4.1   Construction of the Standard Model Lagrangian

Technically, quantum field theory provides the mathematical framework for the Standard Model, in which a Lagrangian controls the dynamics and kinematics of the theory. Each kind of particle is described in terms of a dynamical field that pervades space-time. The construction of the Standard Model proceeds following the modern method of constructing most field theories: by first postulating a set of symmetries of the system, and then by writing down the most general renormalizable Lagrangian from its particle (field) content that observes these symmetries.

The global Poincaré symmetry is postulated for all relativistic quantum field theories. It consists of the familiar translational symmetry, rotational symmetry and the inertial reference frame invariance central to the theory of special relativity. The local SU(3)×SU(2)×U(1) gauge symmetry is an internal symmetry that essentially defines the Standard Model. Roughly, the three factors of the gauge symmetry give rise to the three fundamental interactions. The fields fall into different representations of the various symmetry groups of the Standard Model (see table). Upon writing the most general Lagrangian, one finds that the dynamics depend on 19 parameters, whose numerical values are established by experiment. The parameters are summarized in the table above (note: with the Higgs mass is at 125 GeV, the Higgs self-coupling strength $\lambda \sim 1/8$).

Quantum chromodynamics sector

Main article: Quantum chromodynamics

The quantum chromodynamics (QCD) sector defines the interactions between quarks and gluons, with SU(3) symmetry, generated by T^a. Since leptons do not interact with gluons, they are not affected by this sector. The Dirac Lagrangian of the quarks coupled to the gluon fields is given by

$$\mathcal{L}_{QCD} = i\overline{U}(\partial_\mu - ig_s G_\mu^a T^a)\gamma^\mu U + i\overline{D}(\partial_\mu - ig_s G_\mu^a T^a)\gamma^\mu D.$$

G_μ^a is the SU(3) gauge field containing the gluons, γ^μ are the Dirac matrices, D and U are the Dirac spinors associated with up- and down-type quarks, and g_s is the strong coupling constant.

Electroweak sector

Main article: Electroweak interaction

The electroweak sector is a Yang–Mills gauge theory with the simple symmetry group U(1)×SU(2)L,

$$\mathcal{L}_{EW} = \sum_\psi \bar{\psi}\gamma^\mu \left(i\partial_\mu - g'\frac{1}{2}Y_W B_\mu - g\frac{1}{2}\vec{\tau}_L \vec{W}_\mu \right)\psi$$

where B_μ is the U(1) gauge field; Y_W is the weak hypercharge—the generator of the U(1) group; $\vec{W}_\mu$ is the three-component SU(2) gauge field; $\vec{\tau}_L$ are the Pauli matrices—infinitesimal generators of the SU(2) group. The subscript L indicates that they only act on left fermions; g' and g are coupling constants.

Higgs sector

Main article: Higgs mechanism

In the Standard Model, the Higgs field is a complex scalar of the group SU(2)L:

$$\varphi = \frac{1}{\sqrt{2}}\left(\begin{array}{c} \varphi^+ \\ \varphi^0 \end{array} \right),$$

where the indices + and 0 indicate the electric charge (Q) of the components. The weak isospin (Y_W) of both components is 1.

Before symmetry breaking, the Higgs Lagrangian is:

$$\mathcal{L}_H = \varphi^\dagger \left(\partial^\mu - \frac{i}{2}\left(g'Y_W B^\mu + g\vec{\tau}\vec{W}^\mu \right) \right)\left(\partial_\mu + \frac{i}{2}\left(g'Y_W B_\mu + g\vec{\tau}\vec{W}_\mu \right) \right)\varphi - \frac{\lambda^2}{4}\left(\varphi^\dagger\varphi - v^2 \right)^2,$$

which can also be written as:

$$\mathcal{L}_H = \left| \left(\partial_\mu + \frac{i}{2}\left(g'Y_W B_\mu + g\vec{\tau}\vec{W}_\mu \right) \right)\varphi \right|^2 - \frac{\lambda^2}{4}\left(\varphi^\dagger\varphi - v^2 \right)^2.$$

7.5 Fundamental forces

Main article: Fundamental interaction

The Standard Model classified all four fundamental forces in nature. In the Standard Model, a force is described as an exchange of bosons between the objects affected, such as a photon for the electromagnetic force and a gluon for the strong interaction. Those particles are called force carriers.[26]

7.6   Tests and predictions

The Standard Model (SM) predicted the existence of the W and Z bosons, gluon, and the top and charm quarks before these particles were observed. Their predicted properties were experimentally confirmed with good precision. To give an idea of the success of the SM, the following table compares the measured masses of the W and Z bosons with the masses predicted by the SM:

The SM also makes several predictions about the decay of Z bosons, which have been experimentally confirmed by the Large Electron-Positron Collider at CERN.

In May 2012 BaBar Collaboration reported that their recently analyzed data may suggest possible flaws in the Standard Model of particle physics.[28][29] These data show that a particular type of particle decay called "B to D-star-tau-nu" happens more often than the Standard Model says it should. In this type of decay, a particle called the B-bar meson decays into a D meson, an antineutrino and a tau-lepton. While the level of certainty of the excess (3.4 sigma) is not enough to claim a break from the Standard Model, the results are a potential sign of something amiss and are likely to impact existing theories, including those attempting to deduce the properties of Higgs bosons.[30]

On December 13, 2012, physicists reported the constancy, over space and time, of a basic physical constant of nature that supports the *standard model of physics*. The scientists, studying methanol molecules in a distant galaxy, found the change $(\Delta\mu/\mu)$ in the proton-to-electron mass ratio μ to be equal to "$(0.0 \pm 1.0) \times 10^{-7}$ at redshift z = 0.89" and consistent with "a null result".[31][32]

7.7   Challenges

See also: Physics beyond the Standard Model

Self-consistency of the Standard Model (currently formulated as a non-abelian gauge theory quantized through path-integrals) has not been mathematically proven. While regularized versions useful for approximate computations (for example lattice gauge theory) exist, it is not known whether they converge (in the sense of S-matrix elements) in the limit that the regulator is removed. A key question related to the consistency is the Yang–Mills existence and mass gap problem.

Experiments indicate that neutrinos have mass, which the classic Standard Model did not allow.[33] To accommodate this finding, the classic Standard Model can be modified to include neutrino mass.

If one insists on using only Standard Model particles, this can be achieved by adding a non-renormalizable interaction of leptons with the Higgs boson.[34] On a fundamental level, such an interaction emerges in the seesaw mechanism where heavy right-handed neutrinos are added to the theory. This is natural in the left-right symmetric extension of the Standard Model[35][36] and in certain grand unified theories.[37] As long as new physics appears below or around 10^{14} GeV, the neutrino masses can be of the right order of magnitude.

Theoretical and experimental research has attempted to extend the Standard Model into a Unified field theory or a Theory of everything, a complete theory explaining all physical phenomena including constants. Inadequacies of the Standard Model that motivate such research include:

- It does not attempt to explain gravitation, although a theoretical particle known as a graviton would help explain it, and unlike for the strong and electroweak interactions of the Standard Model, there is no known way of describing general relativity, the canonical theory of gravitation, consistently in terms of quantum field theory. The reason for this is, among other things, that quantum field theories of gravity generally break down before reaching the Planck scale. As a consequence, we have no reliable theory for the very early universe;

- Some consider it to be *ad hoc* and inelegant, requiring 19 numerical constants whose values are unrelated and arbitrary. Although the Standard Model, as it now stands, can explain why neutrinos have masses, the specifics of neutrino mass are still unclear. It is believed that explaining neutrino mass will require an additional 7 or 8 constants, which are also arbitrary parameters;

- The Higgs mechanism gives rise to the hierarchy problem if some new physics (coupled to the Higgs) is present at high energy scales. In these cases in order for the weak scale to be much smaller than the Planck scale, severe fine tuning of the parameters is required; there are, however, other scenarios that include quantum gravity in which such fine tuning can be avoided.[38] There are also issues of Quantum triviality, which suggests that it may not be possible to create a consistent quantum field theory involving elementary scalar particles.

- It should be modified so as to be consistent with the emerging "Standard Model of cosmology." In particular, the Standard Model cannot explain the observed amount of cold dark matter (CDM) and gives contributions to dark energy which are many orders of magnitude too large. It is also difficult to accommodate the observed predominance of matter over antimatter (matter/antimatter asymmetry). The isotropy and homogeneity of the visible universe over large distances seems to require a mechanism like cosmic inflation, which would also constitute an extension of the Standard Model.

- The existence of ultra-high-energy cosmic rays are difficult to explain under the Standard Model.

Currently, no proposed Theory of Everything has been widely accepted or verified.

7.8   See also

- Fundamental interaction:

  - Quantum electrodynamics
  - Strong interaction: Color charge, Quantum chromodynamics, Quark model
  - Weak interaction: Electroweak theory, Fermi theory of beta decay, Weak hypercharge, Weak isospin

- Gauge theory: Nontechnical introduction to gauge theory

- Generation

- Higgs mechanism: Higgs boson, Higgsless model

- J. C. Ward

- J. J. Sakurai Prize for Theoretical Particle Physics

- Lagrangian

- Open questions: BTeV experiment, CP violation, Neutrino masses, Quark matter, Quantum triviality

- Penguin diagram

- Quantum field theory

- Standard Model: Mathematical formulation of, Physics beyond the Standard Model

7.9   Notes and references

[1]   Technically, there are nine such color–anticolor combinations. However, there is one color-symmetric combination that can be constructed out of a linear superposition of the nine combinations, reducing the count to eight.

7.10 References

[1] R. Oerter (2006). *The Theory of Almost Everything: The Standard Model, the Unsung Triumph of Modern Physics* (Kindle ed.). Penguin Group. p. 2. ISBN 0-13-236678-9.

[2] In fact, there are mathematical issues regarding quantum field theories still under debate (see e.g. Landau pole), but the predictions extracted from the Standard Model by current methods applicable to current experiments are all self-consistent. For a further discussion see e.g. Chapter 25 of R. Mann (2010). *An Introduction to Particle Physics and the Standard Model.* CRC Press. ISBN 978-1-4200-8298-2.

[3] Sean Carroll, Ph.D., Cal Tech, 2007, The Teaching Company, *Dark Matter, Dark Energy: The Dark Side of the Universe*, Guidebook Part 2 page 59, Accessed Oct. 7, 2013, "...Standard Model of Particle Physics: The modern theory of elementary particles and their interactions ... It does not, strictly speaking, include gravity, although it's often convenient to include gravitons among the known particles of nature..."

[4] S.L.Glashow(1961). "Partial-symmetries of weak interactions".*Nuclear Physics***22**(4):579–588.Bibcode:1961NucPh..22..5G. doi:10.1016/0029-5582(61)90469-2.

[5] S. Weinberg (1967). "A Model of Leptons". *Physical Review Letters* **19** (21): 1264–1266. Bibcode:1967PhRvL..19.1264W. doi:10.1103/PhysRevLett.19.1264.

[6] A. Salam (1968). N. Svartholm, ed. *Elementary Particle Physics: Relativistic Groups and Analyticity.* Eighth Nobel Symposium. Stockholm: Almquvist and Wiksell. p. 367.

[7] F. Englert, R. Brout (1964). "Broken Symmetry and the Mass of Gauge Vector Mesons". *Physical Review Letters* **13** (9): 321–323. Bibcode:1964PhRvL..13..321E. doi:10.1103/PhysRevLett.13.321.

[8] P.W. Higgs (1964). "Broken Symmetries and the Masses of Gauge Bosons". *Physical Review Letters* **13** (16): 508–509. Bibcode:1964PhRvL..13..508H. doi:10.1103/PhysRevLett.13.508.

[9] G.S. Guralnik, C.R. Hagen, T.W.B. Kibble (1964). "Global Conservation Laws and Massless Particles". *Physical Review Letters* **13** (20): 585–587. Bibcode:1964PhRvL..13..585G. doi:10.1103/PhysRevLett.13.585.

[10] F.J.Hasert et al. (1973). "Search for elastic muon-neutrino electron scattering".*Physics Letters B***46**(1):121.Bibcode:1121H. doi:10.1016/0370-2693(73)90494-2.

[11] F.J. Hasert et al. (1973). "Observation of neutrino-like interactions without muon or electron in the Gargamelle neutrino experiment". *Physics Letters B* **46** (1): 138. Bibcode:1973PhLB...46..138H. doi:10.1016/0370-2693(73)90499-1.

[12] F.J. Hasert et al. (1974). "Observation of neutrino-like interactions without muon or electron in the Gargamelle neutrino experiment". *Nuclear Physics B* **73** (1): 1. Bibcode:1974NuPhB..73....1H. doi:10.1016/0550-3213(74)90038-8.

[13] D. Haidt (4 October 2004). "The discovery of the weak neutral currents". *CERN Courier.* Retrieved 8 May 2008.

[14] "Details can be worked out if the situation is simple enough for us to make an approximation, which is almost never, but often we can understand more or less what is happening." from *The Feynman Lectures on Physics*, Vol 1. pp. 2–7

[15] G.S. Guralnik (2009). "The History of the Guralnik, Hagen and Kibble development of the Theory of Spontaneous Symmetry Breaking and Gauge Particles". *International Journal of Modern Physics A* **24** (14): 2601–2627. arXiv:0907.3466. Bibcode:2009IJMPA..24.2601G. doi:10.1142/S0217751X09045431.

[16] B.W. Lee, C. Quigg, H.B. Thacker (1977). "Weak interactions at very high energies: The role of the Higgs-boson mass". *Physical Review D* **16** (5): 1519–1531. Bibcode:1977PhRvD..16.1519L. doi:10.1103/PhysRevD.16.1519.

[17] "Huge $10 billion collider resumes hunt for 'God particle'". CNN. 11 November 2009. Retrieved 2010-05-04.

[18] M. Strassler (10 July 2012). "Higgs Discovery: Is it a Higgs?". Retrieved 2013-08-06.

[19] "CERN experiments observe particle consistent with long-sought Higgs boson". CERN. 4 July 2012. Retrieved 2012-07-04.

[20] "Observation of a New Particle with a Mass of 125 GeV". CERN. 4 July 2012. Retrieved 2012-07-05.

[21] "ATLAS Experiment". ATLAS. 1 January 2006. Retrieved 2012-07-05.

[22] "Confirmed: CERN discovers new particle likely to be the Higgs boson". *YouTube*. Russia Today. 4 July 2012. Retrieved 2013-08-06.

[23] D. Overbye (4 July 2012). "A New Particle Could Be Physics' Holy Grail". *New York Times*. Retrieved 2012-07-04.

[24] "New results indicate that new particle is a Higgs boson". CERN. 14 March 2013. Retrieved 2013-08-06.

[25] S. Braibant, G. Giacomelli, M. Spurio (2009). *Particles and Fundamental Interactions: An Introduction to Particle Physics*. Springer. pp. 313–314. ISBN 978-94-007-2463-1.

[26] http://home.web.cern.ch/about/physics/standard-model Official CERN website

[27] http://www.pha.jhu.edu/~{}dfehling/particle.gif

[28] "BABAR Data in Tension with the Standard Model". SLAC. 31 May 2012. Retrieved 2013-08-06.

[29] BaBar Collaboration (2012). "Evidence for an excess of B → D$^{(*)}$ τ$^-$ ντ decays". *Physical Review Letters* **109** (10): 101802. arXiv:1205.5442. Bibcode:2012PhRvL.109j1802L. doi:10.1103/PhysRevLett.109.101802.

[30] "BaBar data hint at cracks in the Standard Model". *e! Science News*. 18 June 2012. Retrieved 2013-08-06.

[31] J. Bagdonaite et al. (2012). "A Stringent Limit on a Drifting Proton-to-Electron Mass Ratio from Alcohol in the Early Universe". *Science* **339** (6115): 46. Bibcode:2013Sci...339...46B. doi:10.1126/science.1224898.

[32] C. Moskowitz (13 December 2012). "Phew! Universe's Constant Has Stayed Constant". Space.com. Retrieved 2012-12-14.

[33] "Particle chameleon caught in the act of changing". CERN. 31 May 2010. Retrieved 2012-07-05.

[34] S.Weinberg(1979). "Baryon and Lepton Nonconserving Processes".*Physical Review Letters***43**(21):1566.Bibcode:1979PhRvW. doi:10.1103/PhysRevLett.43.1566.

[35] P.Minkowski(1977). "μ→eγat a Rate of One Out of10₉Muon Decays?".*Physics Letter*1977PhLB...67..421M. doi:10.1016/0370-2693(77)90435-X.

[36] R. N. Mohapatra, G. Senjanovic (1980). "Neutrino Mass and Spontaneous Parity Nonconservation". *Physical Review Letters* **44** (14): 912–915. Bibcode:1980PhRvL..44..912M. doi:10.1103/PhysRevLett.44.912.

[37] M. Gell-Mann, P. Ramond and R. Slansky (1979). F. van Nieuwenhuizen and D. Z. Freedman, ed. *Supergravity*. North Holland. pp. 315–321. ISBN 0-444-85438-X.

[38] Salvio,Strumia(2014-03-17)."Agravity".*JHEP1406(2014)080*.arXiv:1403.4226.Bibcode:2014JHEP...06..080S.doi:1014)080.

7.11 Further reading

- R. Oerter (2006). *The Theory of Almost Everything: The Standard Model, the Unsung Triumph of Modern Physics*. Plume.

- B.A. Schumm (2004). *Deep Down Things: The Breathtaking Beauty of Particle Physics*. Johns Hopkins University Press. ISBN 0-8018-7971-X.

- "The Standard Model of Particle Physics Interactive Graphic".

Introductory textbooks

- I. Aitchison, A. Hey (2003). *Gauge Theories in Particle Physics: A Practical Introduction*. Institute of Physics. ISBN 978-0-585-44550-2.

- W. Greiner, B. Müller (2000). *Gauge Theory of Weak Interactions*. Springer. ISBN 3-540-67672-4.

- G.D. Coughlan, J.E. Dodd, B.M. Gripaios (2006). *The Ideas of Particle Physics: An Introduction for Scientists*. Cambridge University Press.

- D.J. Griffiths (1987). *Introduction to Elementary Particles.* John Wiley & Sons. ISBN 0-471-60386-4.

- G.L. Kane (1987). *Modern Elementary Particle Physics.* Perseus Books. ISBN 0-201-11749-5.

Advanced textbooks

- T.P. Cheng, L.F. Li (2006). *Gauge theory of elementary particle physics.* Oxford University Press. ISBN 0-19-851961-3. Highlights the gauge theory aspects of the Standard Model.

- J.F. Donoghue, E. Golowich, B.R. Holstein (1994). *Dynamics of the Standard Model.* Cambridge University Press. ISBN 978-0-521-47652-2. Highlights dynamical and phenomenological aspects of the Standard Model.

- L. O'Raifeartaigh (1988). *Group structure of gauge theories.* Cambridge University Press. ISBN 0-521-34785-8.

- Nagashima Y. Elementary Particle Physics: Foundations of the Standard Model, Volume 2. (Wiley 2013) 920 рапуы

- Schwartz, M.D. Quantum Field Theory and the Standard Model (Cambridge University Press 2013) 952 pages

- Langacker P. The standard model and beyond. (CRC Press, 2010) 670 pages Highlights group-theoretical aspects of the Standard Model.

Journal articles

- E.S.Abers,B.W.Lee(1973). "Gauge theories".*Physics Reports***9**:1–141.Bibcode:1973PhR.....9....1A.doi:10.10-1573(73)90027-6.

- M. Baak et al. (2012). "The Electroweak Fit of the Standard Model after the Discovery of a New Boson at the LHC". *The European Physical Journal C* **72** (11). arXiv:1209.2716. Bibcode:2012EPJC...72.2205B. doi:10.1140/epjc/s10052-012-2205-9.

- Y. Hayato et al. (1999). "Search for Proton Decay through $p \to \nu K^+$ in a Large Water Cherenkov Detector". *Physical Review Letters***83**(8):1529.arXiv:hep-ex/9904020.Bibcode:1999PhRvL..83.1529H.doi:10.1103/PhysRev9.

- S.F. Novaes (2000). "Standard Model: An Introduction". arXiv:hep-ph/0001283 [hep-ph].

- D.P. Roy (1999). "Basic Constituents of Matter and their Interactions — A Progress Report". arXiv:hep-ph/9912523 [hep-ph].

- F. Wilczek (2004). "The Universe Is A Strange Place". *Nuclear Physics B - Proceedings Supplements* **134**: 3. arXiv:astro-ph/0401347. Bibcode:2004NuPhS.134....3W. doi:10.1016/j.nuclphysbps.2004.08.001.

7.12   External links

- "The Standard Model explained in Detail by CERN's John Ellis" omega tau podcast.

- "LHC sees hint of lightweight Higgs boson" "New Scientist".

- "Standard Model may be found incomplete," *New Scientist.*

- "Observation of the Top Quark" at Fermilab.

- "The Standard Model Lagrangian." After electroweak symmetry breaking, with no explicit Higgs boson.

- "Standard Model Lagrangian" with explicit Higgs terms. PDF, PostScript, and LaTeX versions.

- "The particle adventure." Web tutorial.

- Nobes, Matthew (2002) "Introduction to the Standard Model of Particle Physics" on Kuro5hin: Part 1, Part 2, Part 3a, Part 3b.

- "The Standard Model" The Standard Model on the CERN web site explains how the basic building blocks of matter interact, governed by four fundamental forces.

Chapter 8

Semi-empirical mass formula

In nuclear physics, the **semi-empirical mass formula** (**SEMF**) (sometimes also called **Weizsäcker's formula**, or the **Bethe–Weizsäcker formula**, or the **Bethe–Weizsäcker mass formula** to distinguish it from the Bethe–Weizsäcker process) is used to approximate the mass and various other properties of an atomic nucleus from its number of protons and neutrons. As the name suggests, it is based partly on theory and partly on empirical measurements. The theory is based on the **liquid drop model** proposed by George Gamow, which can account for most of the terms in the formula and gives rough estimates for the values of the coefficients. It was first formulated in 1935 by German physicist Carl Friedrich von Weizsäcker, and although refinements have been made to the coefficients over the years, the structure of the formula remains the same today.[1][2]

The SEMF gives a good approximation for atomic masses and several other effects, but does not explain the appearance of magic numbers of protons and neutrons, and the extra binding-energy and measure of stability that are associated with these numbers of nucleons.

8.1   The liquid drop model and its analysis

The liquid drop model in nuclear physics treats the nucleus as a drop of incompressible nuclear fluid. It was first proposed by George Gamow and then developed by Niels Bohr and John Archibald Wheeler. The fluid is made of nucleons (protons and neutrons), which are held together by the strong nuclear force. This is a crude model that does not explain all the properties of the nucleus, but does explain the spherical shape of most nuclei. It also helps to predict the nuclear binding energy and to assess how much is available for consumption.

Mathematical analysis of the theory delivers an equation which attempts to predict the binding energy of a nucleus in terms of the numbers of protons and neutrons it contains. This equation has five terms on its right hand side. These correspond to the cohesive binding of all the nucleons by the strong nuclear force, a surface energy term, the electrostatic mutual repulsion of the protons, an asymmetry term (derivable from the protons and neutrons occupying independent quantum momentum states) and a pairing term (partly derivable from the protons and neutrons occupying independent quantum spin states).

If we consider the sum of the following five types of energies, then the picture of a nucleus as a drop of incompressible liquid roughly accounts for the observed variation of binding energy of the nucleus:

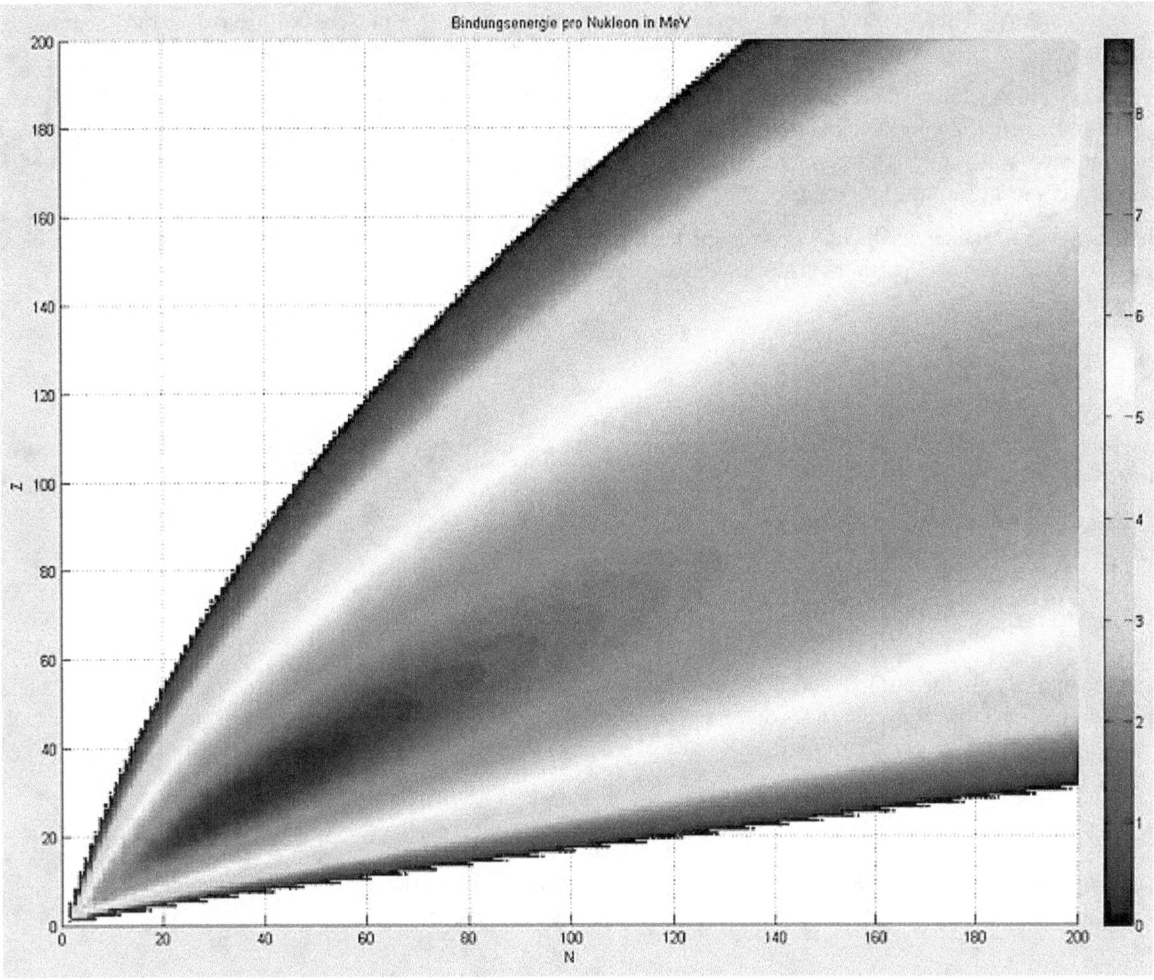

A graphical representation of the semi-empirical binding energy formula. The binding energy per nucleon in MeV (highest numbers in dark red, in excess of 8.5 MeV per nucleon) is plotted for various nuclides as a function of Z, the atomic number (on the y-axis), vs. N, the neutron number (on the x-axis). The highest binding energies are seen for Z = 26 (iron).

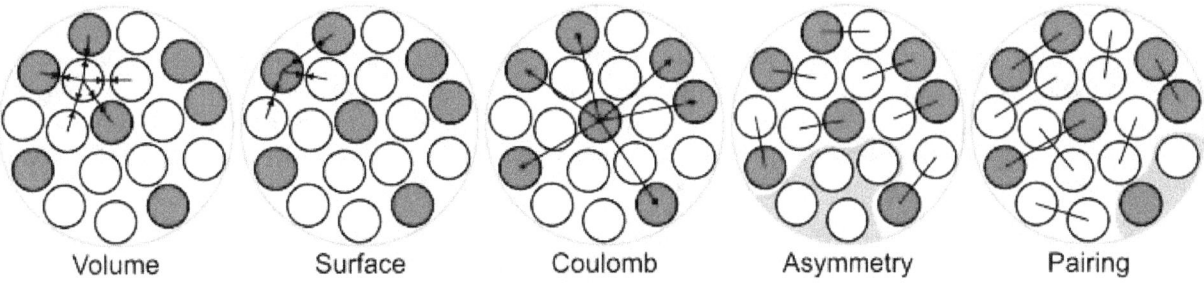

Volume energy. When an assembly of nucleons of the same size is packed together into the smallest volume, each interior nucleon has a certain number of other nucleons in contact with it. So, this nuclear energy is proportional to the volume.

Surface energy. A nucleon at the surface of a nucleus interacts with fewer other nucleons than one in the interior of the nucleus and hence its binding energy is less. This surface energy term takes that into account and is therefore negative and is proportional to the surface area.

Coulomb Energy. The electric repulsion between each pair of protons in a nucleus contributes toward decreasing its binding energy.

Asymmetry energy (also called Pauli Energy). An energy associated with the Pauli exclusion principle. Were it not for the Coulomb energy, the most stable form of nuclear matter would have the same number of neutrons as protons, since unequal numbers of neutrons and protons imply filling higher energy levels for one type of particle, while leaving lower energy levels vacant for the other type.

Pairing energy. An energy which is a correction term that arises from the tendency of proton pairs and neutron pairs to occur. An even number of particles is more stable than an odd number.

8.2 The formula

In the following formulae, let A be the total number of nucleons, Z the number of protons, and N the number of neutrons, so that $A=Z+N$.

The mass of an atomic nucleus is given by

$$m = Zm_p + Nm_n - \frac{E_B}{c^2}$$

where m_p and m_n are the rest mass of a proton and a neutron, respectively, and E_B is the binding energy of the nucleus. The semi-empirical mass formula states that the binding energy will take the following form:

$$E_B = a_V A - a_S A^{2/3} - a_C \frac{Z^2}{A^{1/3}} - a_A \frac{(A - 2Z)^2}{A} - \delta(A, Z)$$

Each of the terms in this formula has a theoretical basis, as will be explained below. The coefficients a_V, a_S, a_C, a_A and a coefficient that appears in the formula for $\delta(A, Z)$ are determined empirically.

8.3 Terms

8.3.1 Volume term

The term $a_V A$ is known as the *volume term*. The volume of the nucleus is proportional to A, so this term is proportional to the volume, hence the name.

The basis for this term is the strong nuclear force. The strong force affects both protons and neutrons, and as expected, this term is independent of Z. Because the number of pairs that can be taken from A particles is $\frac{A(A-1)}{2}$, one might expect a term proportional to A^2. However, the strong force has a very limited range, and a given nucleon may only interact strongly with its nearest neighbors and next nearest neighbors. Therefore, the number of pairs of particles that actually interact is roughly proportional to A, giving the volume term its form.

The coefficient a_V is smaller than the binding energy of the nucleons to their neighbours E_b, which is of order of 40 MeV. This is because the larger the number of nucleons in the nucleus, the larger their kinetic energy is, due to the Pauli exclusion principle. If one treats the nucleus as a Fermi ball of A nucleons, with equal numbers of protons and neutrons, then the total kinetic energy is $\frac{3}{5}A\epsilon_F$, with ϵ_F the Fermi energy which is estimated as 28 MeV. Thus the expected value of a_V in this model is $E_b - \frac{3}{5}\epsilon_F \sim 17$ MeV, not far from the measured value.

8.3.2 Surface term

The term $a_S A^{2/3}$ is known as the *surface term*. This term, also based on the strong force, is a correction to the volume term.

The volume term suggests that each nucleon interacts with a constant number of nucleons, independent of A. While this is very nearly true for nucleons deep within the nucleus, those nucleons on the surface of the nucleus have fewer nearest neighbors, justifying this correction. This can also be thought of as a surface tension term, and indeed a similar mechanism creates surface tension in liquids.

If the volume of the nucleus is proportional to A, then the radius should be proportional to $A^{1/3}$ and the surface area to $A^{2/3}$. This explains why the surface term is proportional to $A^{2/3}$. It can also be deduced that a_S should have a similar order of magnitude as a_V.

8.3.3 Coulomb term

The term $a_C \frac{Z(Z-1)}{A^{1/3}}$ or $a_C \frac{Z^2}{A^{1/3}}$ is known as the *Coulomb* or *electrostatic term*.

The basis for this term is the electrostatic repulsion between protons. To a very rough approximation, the nucleus can be considered a sphere of uniform charge density. The potential energy of such a charge distribution can be shown to be

$$E = \frac{3}{5} \left(\frac{1}{4\pi\epsilon_0} \right) \frac{Q^2}{R}$$

where Q is the total charge and R is the radius of the sphere. Identifying Q with Ze, and noting as above that the radius is proportional to $A^{1/3}$, we get close to the form of the Coulomb term. However, because electrostatic repulsion will only exist for more than one proton, Z^2 becomes $Z(Z-1)$. The value of a_C can be approximately calculated using the equation above:

Empirical nuclear radius:

$$R \approx r_0 A^{\frac{1}{3}}.$$

Quantum charge integers:

$$Q = Ze$$

$$Z^2 \approx Z(Z-1).$$

Integration by substitution:

$$E = \frac{3}{5} \left(\frac{1}{4\pi\epsilon_0} \right) \frac{Q^2}{R} = \frac{3}{5} \left(\frac{1}{4\pi\epsilon_0} \right) \frac{(Ze)^2}{(r_0 A^{\frac{1}{3}})} = \frac{3e^2 Z^2}{20\pi\epsilon_0 r_0 A^{\frac{1}{3}}} \approx \frac{3e^2 Z(Z-1)}{20\pi\epsilon_0 r_0 A^{\frac{1}{3}}} = a_C \frac{Z(Z-1)}{A^{1/3}}$$

Potential energy of charge distribution:

$$E = \frac{3e^2 Z(Z-1)}{20\pi\epsilon_0 r_0 A^{\frac{1}{3}}}$$

Electrostatic Coulomb constant:

$$a_C = \frac{3e^2}{20\pi\epsilon_0 r_0}$$

The value of a_C using the fine structure constant:

$$a_C = \frac{3}{5}\left(\frac{\hbar c \alpha}{r_0}\right) = \frac{3}{5}\left(\frac{R_P}{r_0}\right)\alpha m_p c^2$$

where α is the fine structure constant and $r_0 A^{1/3}$ is the radius of a nucleus, giving r_0 to be approximately 1.25 femtometers. R_P is the proton Compton radius and m_p the proton mass. This gives a_C an approximate theoretical value of 0.691 MeV, not far from the measured value.

$$a_C = 0.691 \text{ MeV}$$

8.3.4 Asymmetry term

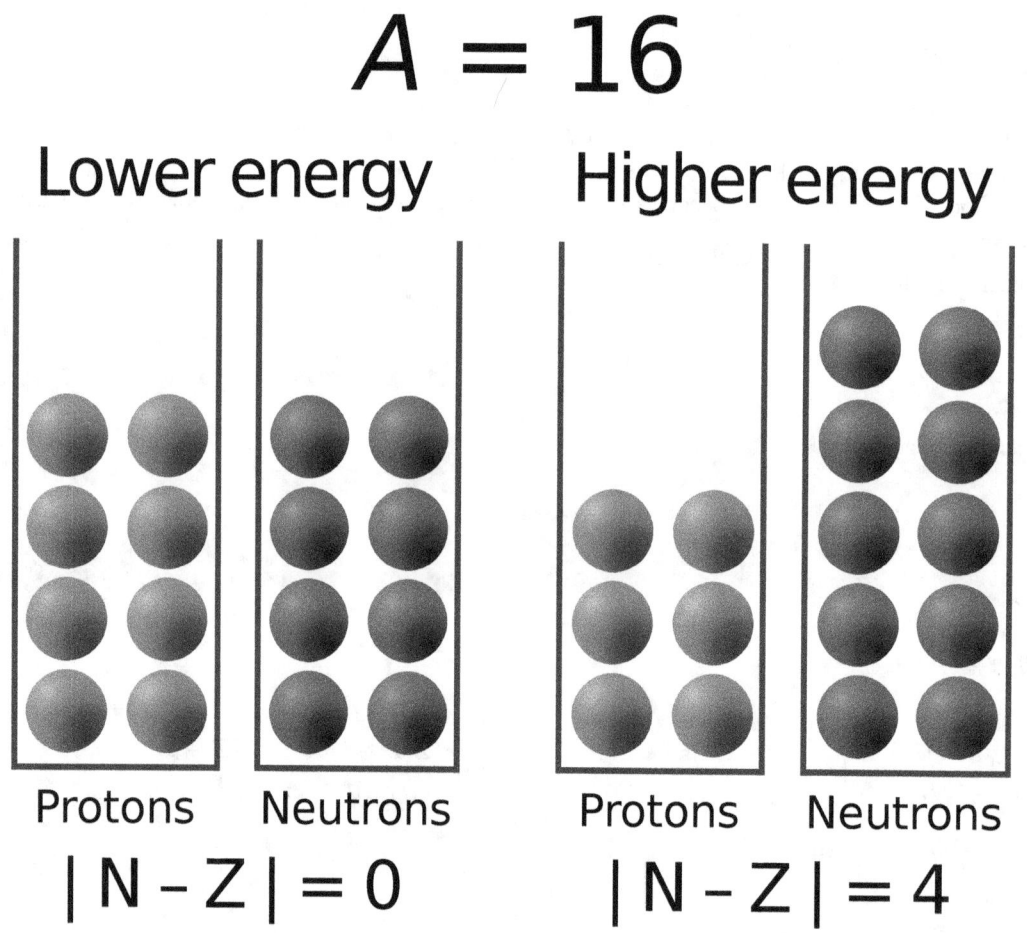

Illustration of basis for asymmetric term

The term $a_A \frac{(A-2Z)^2}{A}$ or $4a_A \frac{((A/2)-Z)^2}{A}$ is known as the *asymmetry term*. Note that as $A = N + Z$, the parenthesized expression can be rewritten as $(N - Z)$. The form $(A - 2Z)$ is used to keep the dependence on A explicit, as will be important for a number of uses of the formula.

The theoretical justification for this term is more complex. The Pauli exclusion principle states that no two fermions can occupy exactly the same quantum state in an atom. At a given energy level, there are only finitely many quantum states available for particles. What this means in the nucleus is that as more particles are "added", these particles must occupy higher energy levels, increasing the total energy of the nucleus (and decreasing the binding energy). Note that this effect is not based on any of the fundamental forces (gravitational, electromagnetic, etc.), only the Pauli exclusion principle.

Protons and neutrons, being distinct types of particles, occupy different quantum states. One can think of two different "pools" of states, one for protons and one for neutrons. Now, for example, if there are significantly more neutrons than protons in a nucleus, some of the neutrons will be higher in energy than the available states in the proton pool. If we could move some particles from the neutron pool to the proton pool, in other words change some neutrons into protons, we would significantly decrease the energy. The imbalance between the number of protons and neutrons causes the energy to be higher than it needs to be, *for a given number of nucleons*. This is the basis for the asymmetry term.

The actual form of the asymmetry term can again be derived by modelling the nucleus as a Fermi ball of protons and neutrons. Its total kinetic energy is

$$E_k = \frac{3}{5}(N_p \epsilon_{Fp} + N_n \epsilon_{Fn})$$

where N_p, N_n are the numbers of protons and neutrons and ϵ_{Fp}, ϵ_{Fn} are their Fermi energies. Since the latter are proportional to $N_p^{2/3}$ and $N_n^{2/3}$, respectively, one gets

$$E_k = C(N_p^{5/3} + N_n^{5/3}) \text{ for some constant } C.$$

The leading expansion in the difference $N_n - N_p$ is then

$$E_k = \frac{C}{2^{2/3}} \left((N_p + N_n)^{5/3} + \frac{5}{9} \frac{(N_n - N_p)^2}{(N_p + N_n)^{1/3}} \right) + O((N_n - N_p)^2).$$

At the zeroth order expansion the kinetic energy is just the Fermi energy $\epsilon_F \equiv \epsilon_{Fp} = \epsilon_{Fn}$ multiplied by $\frac{3}{5}(N_p + N_n)^{2/3}$. Thus we get

$$E_k = \frac{3}{5}\epsilon_F(N_p + N_n) + \frac{1}{3}\epsilon_F \frac{(N_n - N_p)^2}{(N_p + N_n)} + O((N_n - N_p)^4) = \frac{3}{5}\epsilon_F A + \frac{1}{3}\epsilon_F \frac{(A - 2Z)^2}{A} + O((A - 2Z)^4).$$

The first term contributes to the volume term in the semi-empirical mass formula, and the second term is minus the asymmetry term (remember the kinetic energy contributes to the total binding energy with a *negative* sign).

ϵ_F is 38 MeV, so calculating a_A from the equation above, we get only half the measured value. The discrepancy is explained by our model not being accurate: nucleons in fact interact with each other, and are not spread evenly across the nucleus. For example, in the shell model, a proton and a neutron with overlapping wavefunctions will have a greater strong interaction between them and stronger binding energy. This makes it energetically favourable (i.e. having lower energy) for protons and neutrons to have the same quantum numbers (other than isospin), and thus increase the energy cost of asymmetry between them.

One can also understand the asymmetry term intuitively, as follows. It should be dependent on the absolute difference $|N - Z|$, and the form $(A - 2Z)^2$ is simple and differentiable, which is important for certain applications of the formula. In addition, small differences between Z and N do not have a high energy cost. The A in the denominator reflects the fact that a given difference $|N - Z|$ is less significant for larger values of A.

8.3.5   Pairing term

The term $\delta(A, Z)$ is known as the *pairing term* (possibly also known as the pairwise interaction). This term captures the effect of spin-coupling. It is given by:[3]

$$\delta(A,Z) = \begin{cases} +\delta_0 & Z, N \text{ even } (A \text{ even}) \\ 0 & A \text{ odd} \\ -\delta_0 & Z, N \text{ odd } (A \text{ even}) \end{cases}$$

where

$$\delta_0 = \frac{a_P}{A^{1/2}}.$$

Due to Pauli exclusion principle the nucleus would have a lower energy if the number of protons with spin up were equal to the number of protons with spin down. This is also true for neutrons. Only if both Z and N are even can both protons and neutrons have equal numbers of spin up and spin down particles. This is a similar effect to the asymmetry term.

The factor $A^{-1/2}$ is not easily explained theoretically. The Fermi ball calculation we have used above, based on the liquid drop model but neglecting interactions, will give an A^{-1} dependence, as in the asymmetry term. This means that the actual effect for large nuclei will be larger than expected by that model. This should be explained by the interactions between nucleons; For example, in the shell model, two protons with the same quantum numbers (other than spin) will have completely overlapping wavefunctions and will thus have greater strong interaction between them and stronger binding energy. This makes it energetically favourable (i.e. having lower energy) for protons to pair in pairs of opposite spin. The same is true for neutrons.

8.4 Calculating the coefficients

The coefficients are calculated by fitting to experimentally measured masses of nuclei. Their values can vary depending on how they are fitted to the data. Several examples are as shown below, with units of megaelectronvolts.

The semi-empirical mass formula provides a good fit to heavier nuclei, and a poor fit to very light nuclei, especially ^{4}He. This is because the formula does not consider the internal shell structure of the nucleus. For light nuclei, it is usually better to use a model that takes this structure into account.

8.5 Examples for consequences of the formula

By maximizing $B(A,Z)$ with respect to Z, we find the best neutrons to protons ratio N/Z for a given atomic weight A.[5] We get

$$N/Z \approx 1 + \frac{a_C}{2a_A} A^{2/3}.$$

This is roughly 1 for light nuclei, but for heavy nuclei the ratio grows in good agreement with nature.

By substituting the above value of Z back into B one obtains the binding energy as a function of the atomic weight, $B(A)$. Maximizing $B(A)/A$ with respect to A gives the nucleus which is most strongly bound, i.e. most stable. The value we get is $A = 63$ (copper), close to the measured values of $A = 62$ (nickel) and $A = 58$ (iron).

8.6 Notes

[1] von Weizsäcker, C. F. (1935). "Zur Theorie der Kernmassen". *Zeitschrift für Physik* (in German) **96** (7–8): 431–458. Bibcode:1935ZPhy...96..431W. doi:10.1007/BF01337700.

[2] Bailey, D. "Semi-empirical Nuclear Mass Formula". *PHY357: Strings & Binding Energy*. University of Toronto. Retrieved 2011-03-31.

[3] Krane, K. (1988). *Introductory Nuclear Physics*. John Wiley & Sons. p. 68. ISBN 0-471-85914-1.

[4] Wapstra, A.H. (1958). "Atomic Masses of Nuclides". *External Properties of Atomic Nuclei*. Springer. pp. 1–37. doi:10.1007/9-3-642-45901-6_1. ISBN 978-3-642-45902-3.

[5] Rohlf, J. W. (1994). *Modern Physics from α to Z⁰*. John Wiley & Sons. ISBN 978-0471572701.

8.7   References

- Freedman, R.; Young, H. (2004). *Sears and Zemanskey's University Physics with Modern Physics* (11th ed.). pp. 1633–1634. ISBN 0-8053-8768-4.

- Liverhant, S. E. (1960). *Elementary Introduction to Nuclear Reactor Physics*. John Wiley & Sons. pp. 58–62. LCCN 60011725.

- Choppin, G.; Liljenzin, J.-O.; Rydberg, J. (2002). "Nuclear Mass and Stability" (PDF). *Radiochemistry and Nuclear Chemistry* (3rd ed.). Butterworth-Heinemann. pp. 41–57. ISBN 978-0-7506-7463-8.

8.8   External links

- Nuclear liquid drop model

- The semi-empirical mass formula

- Liquid drop model in the hyperphysics online reference at Georgia State University.

- Liquid drop model with parameter fit from *First Observations of Excited States in the Neutron Deficient Nuclei $^{160,161}W$ and ^{159}Ta*, Alex Keenan, PhD thesis, University of Liverpool, 1999 (HTML version).

Chapter 9

Nuclear shell model

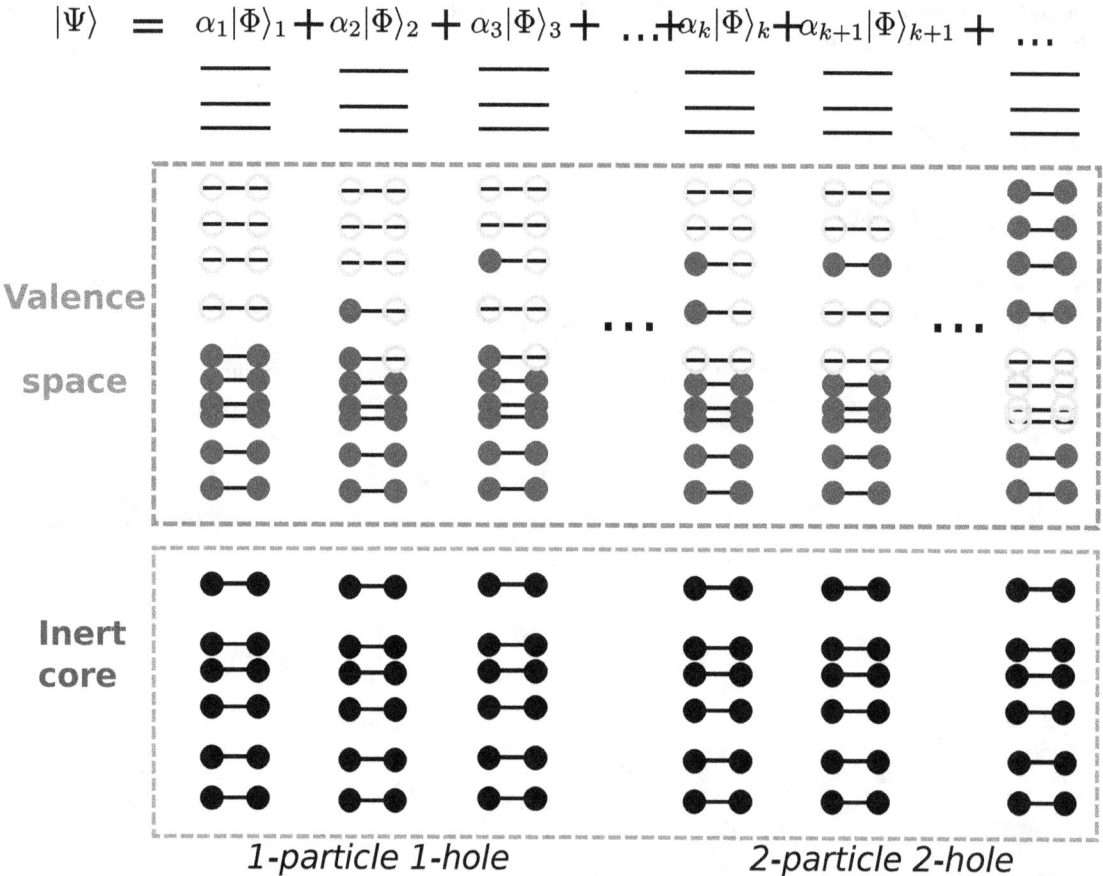

$$|\Psi\rangle \; = \; \alpha_1|\Phi\rangle_1 + \alpha_2|\Phi\rangle_2 + \alpha_3|\Phi\rangle_3 + \ldots + \alpha_k|\Phi\rangle_k + \alpha_{k+1}|\Phi\rangle_{k+1} + \ldots$$

1-particle 1-hole   *2-particle 2-hole*

Partly filled valence orbitals for both neutrons and protons appear at energies over the filled inert core orbitals, in the shell model of the atomic nucleus

In nuclear physics and nuclear chemistry, the **nuclear shell model** is a model of the atomic nucleus which uses the Pauli exclusion principle to describe the structure of the nucleus in terms of energy levels.[1] The first shell model was proposed by Dmitry Ivanenko (together with E. Gapon) in 1932. The model was developed in 1949 following independent work by several physicists, most notably Eugene Paul Wigner, Maria Goeppert Mayer and J. Hans D. Jensen, who shared the 1963 Nobel Prize in Physics for their contributions.

The shell model is partly analogous to the atomic shell model which describes the arrangement of electrons in an atom, in that a filled shell results in greater stability. When adding nucleons (protons or neutrons) to a nucleus, there are certain points where the binding energy of the next nucleon is significantly less than the last one. This observation, that there are certain magic numbers of nucleons: 2, 8, 20, 28, 50, 82, 126 which are more tightly bound than the next higher number, is the origin of the shell model.

The shells for protons and for neutrons are independent of each other. Therefore, one can have "magic nuclei" where one nucleon type or the other is at a magic number, and "doubly magic nuclei", where both are. Due to some variations in orbital filling, the upper magic numbers are 126 and, speculatively, 184 for neutrons but only 114 for protons, playing a role in the search for the so-called island of stability. Some semimagic numbers have been found, notably Z=40 giving nuclear shell filling for the various elements; 16 may also be a magic number.[2]

In order to get these numbers, the nuclear shell model starts from an average potential with a shape something between the square well and the harmonic oscillator. To this potential a spin orbit term is added. Even so, the total perturbation does not coincide with experiment, and an empirical spin orbit coupling, named the Nilsson Term, must be added with at least two or three different values of its coupling constant, depending on the nuclei being studied.

Nevertheless, the magic numbers of nucleons, as well as other properties, can be arrived at by approximating the model with a three-dimensional harmonic oscillator plus a spin-orbit interaction. A more realistic but also complicated potential is known as Woods Saxon potential.

Igal Talmi developed a method to obtain the information from experimental data and use it to calculate and predict energies which have not been measured. This method has been successfully used by many nuclear physicists and has led to deeper understanding of nuclear structure. The theory which gives a good description of these properties was developed. This description turned out to furnish the shell model basis of the elegant and successful Interacting boson model.

9.1 Deformed harmonic oscillator approximated model

Consider a three-dimensional harmonic oscillator. This would give, for example, in the first two levels ("l" is angular momentum)

We can imagine ourselves building a nucleus by adding protons and neutrons. These will always fill the lowest available level. Thus the first two protons fill level zero, the next six protons fill level one, and so on. As with electrons in the periodic table, protons in the outermost shell will be relatively loosely bound to the nucleus if there are only few protons in that shell, because they are farthest from the center of the nucleus. Therefore nuclei which have a full outer proton shell will have a higher binding energy than other nuclei with a similar total number of protons. All this is true for neutrons as well.

This means that the magic numbers are expected to be those in which all occupied shells are full. We see that for the first two numbers we get 2 (level 0 full) and 8 (levels 0 and 1 full), in accord with experiment. However the full set of magic numbers does not turn out correctly. These can be computed as follows:

In a three-dimensional harmonic oscillator the total degeneracy at level n is $\frac{(n+1)(n+2)}{2}$. Due to the spin, the degeneracy is doubled and is $(n+1)(n+2)$.

Thus the magic numbers would be

$$\sum_{n=0}^{k}(n+1)(n+2) = \frac{(k+1)(k+2)(k+3)}{3}$$

for all integer k. This gives the following magic numbers: 2,8,20,40,70,112..., which agree with experiment only in the first three entries. These numbers are twice the tetrahedral numbers (1,4,10,20,35,56...) from the Pascal Triangle.

In particular, the first six shells are:

- level 0: 2 states ($l = 0$) = 2.

- level 1: 6 states ($l = 1$) = 6.

- level 2: 2 states ($l = 0$) + 10 states ($l = 2$) = 12.

- level 3: 6 states ($l = 1$) + 14 states ($l = 3$) = 20.

- level 4: 2 states ($l = 0$) + 10 states ($l = 2$) + 18 states ($l = 4$) = 30.

- level 5: 6 states ($l = 1$) + 14 states ($l = 3$) + 22 states ($l = 5$) = 42.

where for every l there are $2l+1$ different values of ml and 2 values of ms, giving a total of $4l+2$ states for every specific level.

These numbers are twice the values of triangular numbers from the Pascal Triangle: 1,3,6,10,15,21....

9.1.1 Including a spin-orbit interaction

We next include a spin-orbit interaction. First we have to describe the system by the quantum numbers j, mj and parity instead of l, ml and ms, as in the Hydrogen-like atom. Since every even level includes only even values of l, it includes only states of even (positive) parity; Similarly every odd level includes only states of odd (negative) parity. Thus we can ignore parity in counting states. The first six shells, described by the new quantum numbers, are

- level 0 ($n=0$): 2 states ($j = \frac{1}{2}$). Even parity.

- level 1 ($n=1$): 2 states ($j = \frac{1}{2}$) + 4 states ($j = \frac{3}{2}$) = 6. Odd parity.

- level 2 ($n=2$): 2 states ($j = \frac{1}{2}$) + 4 states ($j = \frac{3}{2}$) + 6 states ($j = \frac{5}{2}$) = 12. Even parity.

- level 3 ($n=3$): 2 states ($j = \frac{1}{2}$) + 4 states ($j = \frac{3}{2}$) + 6 states ($j = \frac{5}{2}$) + 8 states ($j = \frac{7}{2}$) = 20. Odd parity.

- level 4 ($n=4$): 2 states ($j = \frac{1}{2}$) + 4 states ($j = \frac{3}{2}$) + 6 states ($j = \frac{5}{2}$) + 8 states ($j = \frac{7}{2}$) + 10 states ($j = \frac{9}{2}$) = 30. Even parity.

- level 5 ($n=5$): 2 states ($j = \frac{1}{2}$) + 4 states ($j = \frac{3}{2}$) + 6 states ($j = \frac{5}{2}$) + 8 states ($j = \frac{7}{2}$) + 10 states ($j = \frac{9}{2}$) + 12 states ($j = \frac{11}{2}$) = 42. Odd parity.

where for every j there are $2j+1$ different states from different values of mj.

Due to the spin-orbit interaction the energies of states of the same level but with different j will no longer be identical. This is because in the original quantum numbers, when $\vec{s}$ is parallel to $\vec{l}$, the interaction energy is positive; and in this case $j = l + s = l + \frac{1}{2}$. When $\vec{s}$ is anti-parallel to $\vec{l}$ (i.e. aligned oppositely), the interaction energy is negative, and in this case $j=l-s=l-\frac{1}{2}$. Furthermore, the strength of the interaction is roughly proportional to l.

For example, consider the states at level 4:

- The 10 states with $j = \frac{9}{2}$ come from $l = 4$ and s parallel to l. Thus they have a positive spin-orbit interaction energy.

- The 8 states with $j = \frac{7}{2}$ came from $l = 4$ and s anti-parallel to l. Thus they have a negative spin-orbit interaction energy.

- The 6 states with $j = \frac{5}{2}$ came from $l = 2$ and s parallel to l. Thus they have a positive spin-orbit interaction energy. However its magnitude is half compared to the states with $j = \frac{9}{2}$.

- The 4 states with $j = \frac{3}{2}$ came from $l = 2$ and s anti-parallel to l. Thus they have a negative spin-orbit interaction energy. However its magnitude is half compared to the states with $j = \frac{7}{2}$.

- The 2 states with $j = \frac{1}{2}$ came from $l = 0$ and thus have zero spin-orbit interaction energy.

9.1.2 Deforming the potential

The harmonic oscillator potential $V(r) = \mu\omega^2 r^2/2$ grows infinitely as the distance from the center r goes to infinity. A more realistic potential, such as Woods Saxon potential, would approach a constant at this limit. One main consequence is that the average radius of nucleons' orbits would be larger in a realistic potential; This leads to a reduced term $\hbar^2 l(l+1)/2mr^2$ in the Laplace operator of the Hamiltonian. Another main difference is that orbits with high average radii, such as those with high n or high l, will have a lower energy than in a harmonic oscillator potential. Both effects lead to a reduction in the energy levels of high l orbits.

9.1.3 Predicted magic numbers

Together with the spin-orbit interaction, and for appropriate magnitudes of both effects, one is led to the following qualitative picture: At all levels, the highest j states have their energies shifted downwards, especially for high n (where the highest j is high). This is both due to the negative spin-orbit interaction energy and to the reduction in energy resulting from deforming the potential to a more realistic one. The second-to-highest j states, on the contrary, have their energy shifted up by the first effect and down by the second effect, leading to a small overall shift. The shifts in the energy of the highest j states can thus bring the energy of states of one level to be closer to the energy of states of a lower level. The "shells" of the shell model are then no longer identical to the levels denoted by n, and the magic numbers are changed.

We may then suppose that the highest j states for $n = 3$ have an intermediate energy between the average energies of $n = 2$ and $n = 3$, and suppose that the highest j states for larger n (at least up to $n = 7$) have an energy closer to the average energy of $n-1$. Then we get the following shells (see the figure)

- 1st shell: 2 states ($n = 0$, $j = \frac{1}{2}$).

- 2nd shell: 6 states ($n = 1$, $j = \frac{1}{2}$ or $\frac{3}{2}$).

- 3rd shell: 12 states ($n = 2$, $j = \frac{1}{2}$, $\frac{3}{2}$ or $\frac{5}{2}$).

- 4th shell: 8 states ($n = 3$, $j = \frac{7}{2}$).

- 5th shell: 22 states ($n = 3$, $j = \frac{1}{2}$, $\frac{3}{2}$ or $\frac{5}{2}$; $n = 4$, $j = \frac{9}{2}$).

- 6th shell: 32 states ($n = 4$, $j = \frac{1}{2}$, $\frac{3}{2}$, $\frac{5}{2}$ or $\frac{7}{2}$; $n = 5$, $j = \frac{11}{2}$).

- 7th shell: 44 states ($n = 5$, $j = \frac{1}{2}$, $\frac{3}{2}$, $\frac{5}{2}$, $\frac{7}{2}$ or $\frac{9}{2}$; $n = 6$, $j = \frac{13}{2}$).

- 8th shell: 58 states ($n = 6$, $j = \frac{1}{2}$, $\frac{3}{2}$, $\frac{5}{2}$, $\frac{7}{2}$, $\frac{9}{2}$ or $\frac{11}{2}$; $n = 7$, $j = \frac{15}{2}$).

and so on.

The magic numbers are then

- 2

- 8=2+6

- 20=2+6+12

- 28=2+6+12+8

- 50=2+6+12+8+22

- 82=2+6+12+8+22+32

- 126=2+6+12+8+22+32+44

- 184=2+6+12+8+22+32+44+58

and so on. This gives all the observed magic numbers, and also predicts a new one (the so-called *island of stability*) at the value of 184 (for protons, the magic number 126 has not been observed yet, and more complicated theoretical considerations predict the magic number to be 114 instead).

Another way to predict magic (and semi-magic) numbers is by laying out the idealized filling order (with spin-orbit splitting but energy levels not overlapping). For consistency s is split into $j = 1/2$ and $j = -1/2$ components with 2 and 0 members respectively. Taking leftmost and rightmost total counts within sequences marked bounded by / here gives the magic and semi-magic numbers.

- s(2,0)/p(4,2)> 2,2/6,8, so (semi)magic numbers 2,2/6,8

- d(6,4):s(2,0)/f(8,6):p(4,2)> 14,18:20,20/28,34:38,40, so 14,20/28,40

- g(10,8):d(6,4):s(2,0)/h(12,10):f(8,6):p(4,2)> 50,58,64,68,70,70/82,92,100,106,110,112, so 50,70/82,112

- i(14,12):g(10,8):d(6,4):s(2,0)/j(16,14):h(12,10):f(8,6):p(4,2)> 126,138,148,156,162,166,168,168/184,198, 210,220,228,234,238,240, so 126,168/184,240

The rightmost predicted magic numbers of each pair within the quartets bisected by / are double tetrahedral numbers from the Pascal Triangle: 2,8,20,40,70,112,168,240 are 2x 1,4,10,20,35,56,84,120..., and the leftmost members of the pairs differ from the rightmost by double triangular numbers: 2-2=0, 8-6=2, 20-14=6, 40-28=12, 70-50=20, 112-82=30, 168-126=42, 240-184=56, where 0,2,6,12,20,30,42,56... are 2x 0,1,3,6,10,15,21,28....

9.1.4 Other properties of nuclei

This model also predicts or explains with some success other properties of nuclei, in particular spin and parity of nuclei ground states, and to some extent their excited states as well. Take 17
8O as an example — its nucleus has eight protons filling the three first proton 'shells', eight neutrons filling the three first neutron 'shells', and one extra neutron. All protons in a complete proton shell have total angular momentum zero, since their angular momenta cancel each other; The same is true for neutrons. All protons in the same level (n) have the same parity (either $+1$ or -1), and since the parity of a pair of particles is the product of their parities, an even number of protons from the same level (n) will have $+1$ parity. Thus the total angular momentum of the eight protons and the first eight neutrons is zero, and their total parity is $+1$. This means that the spin (i.e. angular momentum) of the nucleus, as well as its parity, are fully determined by that of the ninth neutron. This one is in the first (i.e. lowest energy) state of the 4th shell, which is a d-shell ($l = 2$), and since $p = (-1)^l$, this gives the nucleus an overall parity of $+1$. This 4th d-shell has a $j = 5/2$, thus the nucleus of 17
8O is expected to have positive parity and total angular momentum $5/2$, which indeed it has.

The rules for the ordering of the nucleus shells are similar to Hund's Rules of the atomic shells, however, unlike its use in atomic physics the completion of a shell is not signified by reaching the next n, as such the shell model cannot accurately predict the order of excited nuclei states, though it is very successful in predicting the ground states. The order of the first few terms are listed as follows: 1s, 1p$3/2$, 1p$1/2$, 1d$5/2$, 2s, 1d$3/2$... For further clarification on the notation refer to the article on the Russell-Saunders term symbol.

For nuclei farther from the magic numbers one must add the assumption that due to the relation between the strong nuclear force and angular momentum, protons or neutrons with the same n tend to form pairs of opposite angular momenta. Therefore a nucleus with an even number of protons and an even number of neutrons has 0 spin and positive parity. A nucleus with an even number of protons and an odd number of neutrons (or vice versa) has the parity of the last neutron (or proton), and the spin equal to the total angular momentum of this neutron (or proton). By "last" we mean the properties coming from the highest energy level.

In the case of a nucleus with an odd number of protons and an odd number of neutrons, one must consider the total angular momentum and parity of both the last neutron and the last proton. The nucleus parity will be a product of theirs, while the nucleus spin will be one of the possible results of the sum of their angular momenta (with other possible results being excited states of the nucleus).

The ordering of angular momentum levels within each shell is according to the principles described above - due to spin-orbit interaction, with high angular momentum states having their energies shifted downwards due to the deformation of the potential (i.e. moving from a harmonic oscillator potential to a more realistic one). For nucleon pairs, however, it is often energetically favorable to be at high angular momentum, even if its energy level for a single nucleon would be higher. This is due to the relation between angular momentum and the strong nuclear force.

Nuclear magnetic moment is partly predicted by this simple version of the shell model. The magnetic moment is calculated through j, l and s of the "last" nucleon, but nuclei are not in states of well defined l and s. Furthermore, for odd-odd nuclei, one has to consider the two "last" nucleons, as in deuterium. Therefore one gets several possible answers for the nuclear magnetic moment, one for each possible combined l and s state, and the real state of the nucleus is a superposition of them. Thus the real (measured) nuclear magnetic moment is somewhere in between the possible answers.

The electric dipole of a nucleus is always zero, because its ground state has a definite parity, so its matter density (ψ^2 , where ψ is the wavefunction) is always invariant under parity. This is usually the situations with the atomic electric dipole as well.

Higher electric and magnetic multipole moments cannot be predicted by this simple version of the shell model, for the reasons similar to those in the case of deuterium.

9.2   Alpha particle model

A model derived from the nuclear shell model is the alpha particle model developed by Henry Margenau, Edward Teller, J. K. Pering, T.H. Skyrme.

9.3   See also

- Interacting boson model

- Liquid drop model

- Nuclear structure

- Isomeric shift

9.4   References

[1] "Shell Model of Nucleus". *HyperPhysics*.

[2] Ozawa, A.; Kobayashi, T.; Suzuki, T.; Yoshida, K.; Tanihata, I. (2000). "New Magic Number, N=16, near the Neutron Drip Line". *Physical Review Letters* **84** (24): 5493–5. Bibcode:2000PhRvL..84.5493O. doi:10.1103/PhysRevLett.84.5493. PMID 10990977. (this refers to the nuclear drip line)

9.5   Books

- Talmi, Igal; de-Shalit, A. (1963). *Nuclear Shell Theory*. Academic Press, (reprinted by Dover Publications). ISBN 0-486-43933-X.

- Talmi, Igal (1993). *Simple Models of Complex Nuclei: The Shell Model and the Interacting Boson Model*. Harwood Academic Publishers. ISBN 3-7186-0551-1.

9.6 Recent Lecture

- Igal Talmi (Nov 24, 2010). *On single nucleon wave functions.* RIKEN Nishina Center.

Internet streaming broadcasting both on WM and QT at (at 64 kbit/s, 256 kbit/s, 1 Mbit/s) and DVD ISO (NTSC and PAL) delivery are now available at RIKEN Nishina Center.

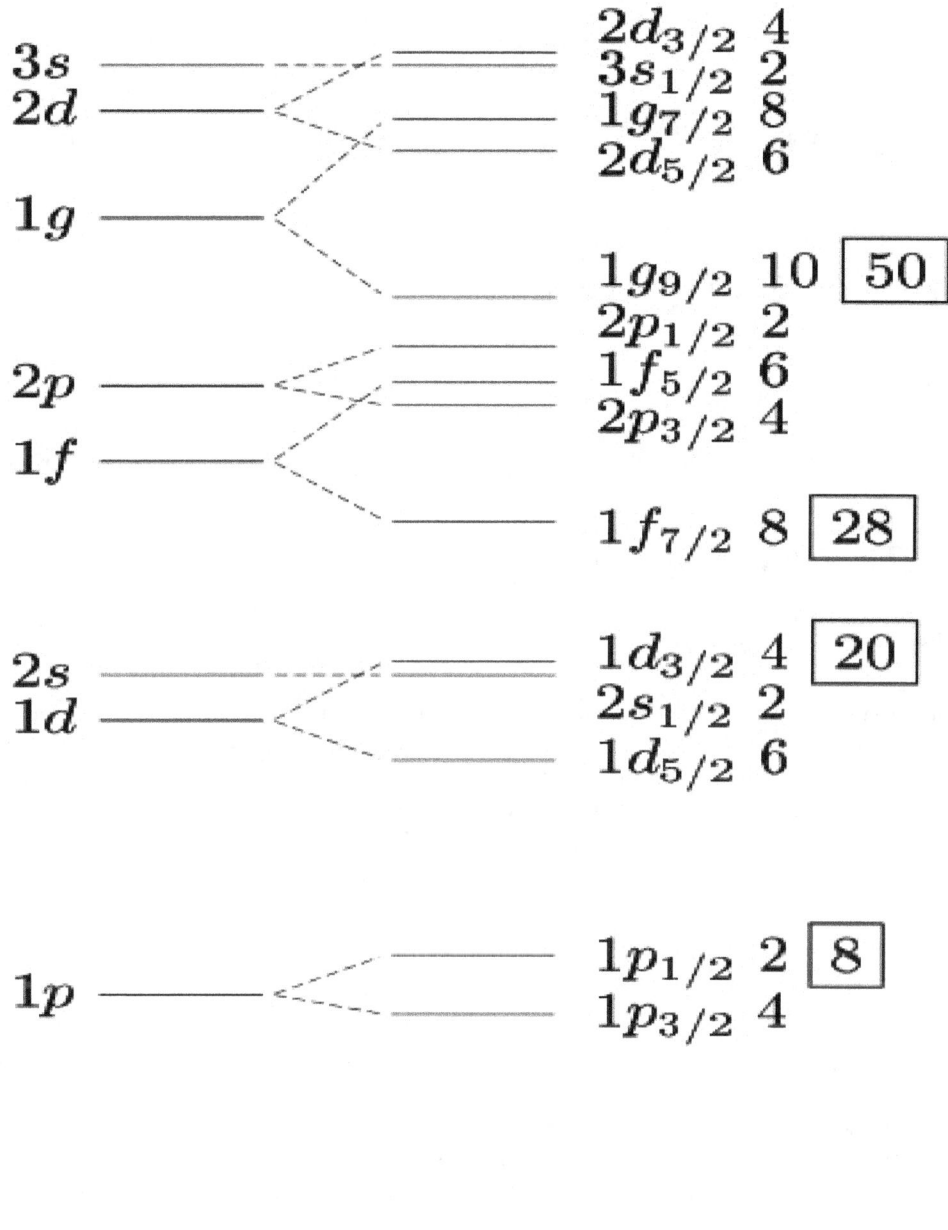

Low-lying energy levels in a single-particle shell model with an oscillator potential (with a small negative l² term) without spin-orbit (left)

Chapter 10

Nuclear structure

Understanding the structure of the atomic nucleus is one of the central challenges in nuclear physics.

10.1 The liquid drop model

Main article: Semi-empirical mass formula

The liquid drop model is one of the first models of **nuclear structure**, proposed by Carl Friedrich von Weizsäcker in 1935.[1] It describes the nucleus as a semiclassical fluid made up of neutrons and protons, with an internal repulsive electrostatic force proportional to the number of protons. The quantum mechanical nature of these particles appears via the Pauli exclusion principle, which states that no two nucleons of the same kind can be at the same state. Thus the fluid is actually what is known as a Fermi liquid. In this model, the binding energy of a nucleus with Z protons and N neutrons is given by

$$E_B = a_V A - a_S A^{2/3} - a_C \frac{Z^2}{A^{1/3}} - a_A \frac{(N - Z)^2}{A} - \delta(A, Z)$$

where $A = Z + N$ is the total number of nucleons. The terms proportional to A and $A^{2/3}$ represent the volume and surface energy of the liquid drop, the term proportional to Z^2 represents the electrostatic energy, the term proportional to $(N - Z)^2$ represents the Pauli exclusion principle and the last term $\delta(A, Z)$ is the pairing term, which lowers the energy for even numbers of protons or neutrons. The coefficients a_V, a_S, a_C, a_A and the strength of the pairing term may be estimated theoretically, or fit to data. This simple model reproduces the main features of the binding energy of nuclei.

The assumption of nucleus as a drop of Fermi liquid is still widely used in the form of Finite Range Droplet Model (FRDM), due to the possible good reproduction of nuclear binding energy on the whole chart, with the necessary accuracy for predictions of unknown nuclei.[2]

10.2 The shell model

Main article: Nuclear shell model

The expression "shell model" is ambiguous in that it refers to two different eras in the state of the art. It was previously used to describe the existence of nucleon shells in the nucleus according to an approach closer to what is now called mean field theory. Nowadays, it refers to a formalism analogous to the configuration interaction formalism used in quantum chemistry. We shall introduce the latter here.

10.2.1   Introduction to the shell concept

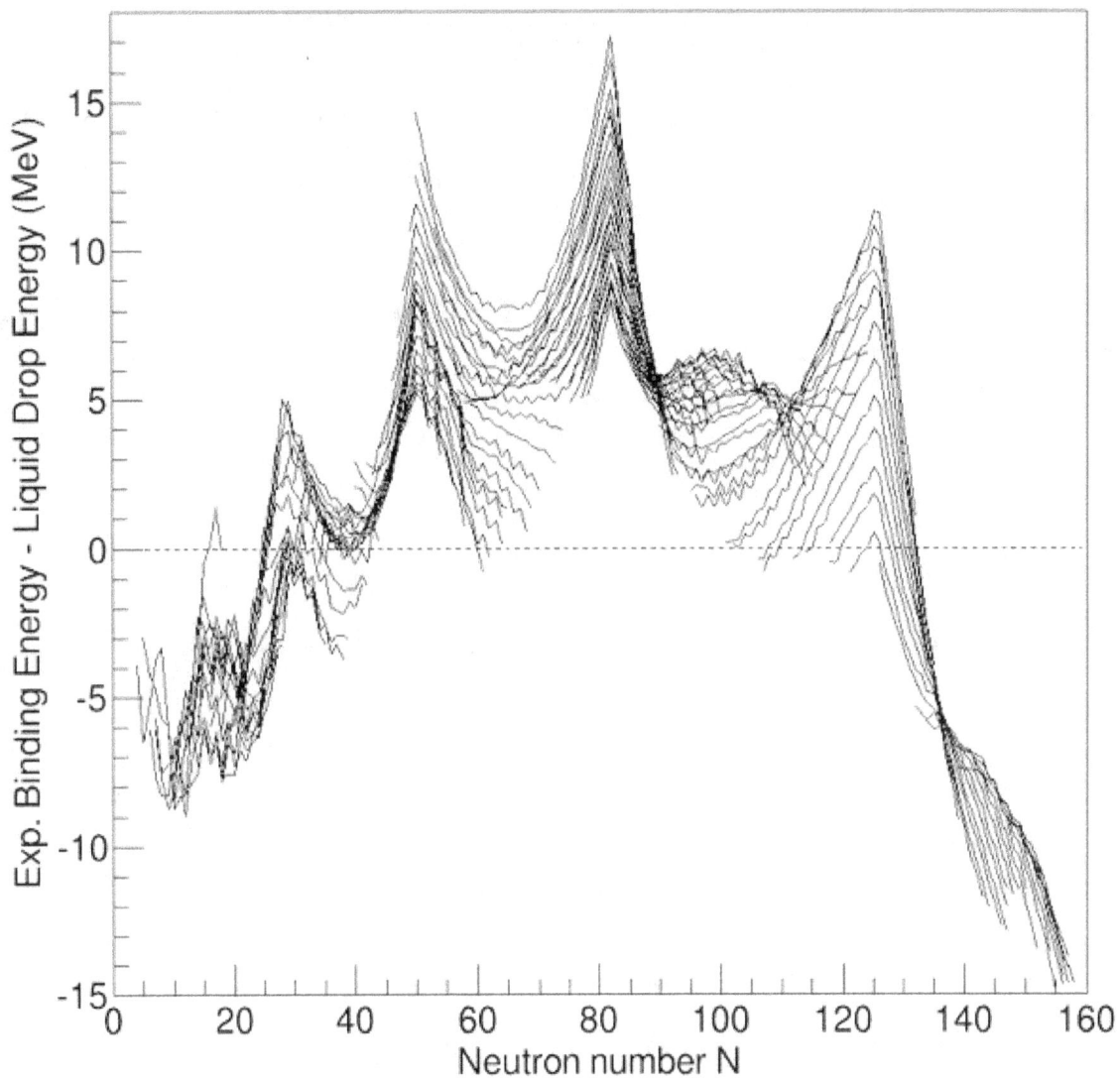

Difference between experimental binding energies and the liquid drop model prediction as a function of neutron number for Z>7

Systematic measurements of the binding energy of atomic nuclei show systematic deviations with respect to those estimated from the liquid drop model. In particular, some nuclei having certain values for the number of protons and/or neutrons are bound more tightly together than predicted by the liquid drop model. These nuclei are called singly/doubly magic. This observation led scientists to assume the existence of a shell structure of nucleons (protons and neutrons) within the nucleus, like that of electrons within atoms.

Indeed, nucleons are quantum objects. Strictly speaking, one should not speak of energies of individual nucleons, because they are all correlated with each other. However, as an approximation one may envision an average nucleus, within which nucleons propagate individually. Owing to their quantum character, they may only occupy *discrete* energy levels. These levels are by no means uniformly distributed; some intervals of energy are crowded, and some are empty, generating a gap in possible energies. A shell is such a set of levels separated from the other ones by a wide empty gap.

The energy levels are found by solving the Schrödinger equation for a single nucleon moving in the average potential generated by all other nucleons. Each level may be occupied by a nucleon, or empty. Some levels accommodate several different quantum states with the same energy; they are said to be *degenerate*. This occurs in particular if the average nucleus has some symmetry.

The concept of shells allows one to understand why some nuclei are bound more tightly than others. This is because two nucleons of the same kind cannot be in the same state (Pauli exclusion principle). So the lowest-energy state of the nucleus is one where nucleons fill all energy levels from the bottom up to some level. A nucleus with full shells is exceptionally stable, as will be explained.

As with electrons in the electron shell model, protons in the outermost shell are relatively loosely bound to the nucleus if there are only few protons in that shell, because they are farthest from the center of the nucleus. Therefore, nuclei which have a full outer proton shell will be more tightly bound and have a higher binding energy than other nuclei with a similar total number of protons. All this is also true for neutrons.

Furthermore, the energy needed to excite the nucleus (i.e. moving a nucleon to a higher, previously unoccupied level) is exceptionally high in such nuclei. Whenever this unoccupied level is the next after a full shell, the only way to excite the nucleus is to raise one nucleon *across the gap*, thus spending a large amount of energy. Otherwise, if the highest occupied energy level lies in a partly filled shell, much less energy is required to raise a nucleon to a higher state in the same shell.

Some evolution of the shell structure observed in stable nuclei is expected away from the valley of stability. For example, observations of unstable isotopes have shown shifting and even a reordering of the single particle levels of which the shell structure is composed.[3] This is sometimes observed as the creation of an island of inversion or in the reduction of excitation energy gaps above the traditional magic numbers.

10.2.2 Basic hypotheses

Some basic hypotheses are made in order to give a precise conceptual framework to the shell model :

- The atomic nucleus is a quantum n-body system.

- The internal motion of nucleons within the nucleus is non-relativistic, and their behavior is governed by the Schrödinger equation.

- Nucleons are considered to be pointlike, without any internal structure.

10.2.3 Brief description of the formalism

The general process used in the shell model calculations is the following. First a Hamiltonian for the nucleus is defined. Usually, for computational practicality, only one- and two-body terms are taken into account in this definition. The interaction is an effective theory : it contains free parameters which have to be fitted with experimental data.

The next step consists in defining a basis of single-particle states, i.e. a set of wavefunctions describing all possible nucleon states. Most of the time, this basis is obtained via a Hartree–Fock computation. With this set of one-particle states, Slater determinants are built, that is, wavefunctions for Z proton variables or N neutron variables, which are antisymmetrized products of single-particle wavefunctions (antisymmetrized meaning that under exchange of variables for any pair of nucleons, the wavefunction only changes sign).

In principle, the number of quantum states available for a single nucleon at a finite energy is finite, say n. The number of nucleons in the nucleus must be smaller than the number of available states, otherwise the nucleus cannot hold all of its nucleons. There are thus several ways to choose Z (or N) states among the n possible. In combinatorial mathematics, the number of choices of Z objects among n is the binomial coefficient C^Z_n. If n is much larger than Z (or N), this increases roughly like n^Z. Practically, this number becomes so large that every computation is impossible for $A=N+Z$ larger than 8.

To obviate this difficulty, the space of possible single-particle states is divided into a core and a valence shell, by analogy with chemistry. The core is a set of single-particles which are assumed to be inactive, in the sense that they are the well bound lowest-energy states, and that there is no need to reexamine their situation. They do not appear in the Slater determinants, contrary to the states in the valence space, which is the space of all single-particle states *not in the core*, but possibly to be considered in the choice of the build of the (Z-) N-body wavefunction. The set of all possible Slater determinants in the valence space defines a basis for (Z-) N-body states.

The last step consists in computing the matrix of the Hamiltonian within this basis, and to diagonalize it. In spite of the reduction of the dimension of the basis owing to the fixation of the core, the matrices to be diagonalized reach easily dimensions of the order of 10^9, and demand specific diagonalization techniques.

The shell model calculations give in general an excellent fit with experimental data. They depend however strongly on two main factors :

- The way to divide the single-particle space into core and valence.

- The effective nucleon–nucleon interaction.

10.3   Mean field theories

10.3.1   The independent-particle model

The interaction between nucleons, which is a consequence of strong interactions and binds the nucleons within the nucleus, exhibits the peculiar behaviour of having a finite range: it vanishes when the distance between two nucleons becomes too large; it is attractive at medium range, and repulsive at very small range. This last property correlates with the Pauli exclusion principle according to which two fermions (nucleons are fermions) cannot be in the same quantum state. This results in a very large mean free path predicted for a nucleon within the nucleus.[4]

The main idea of the Independent Particle approach is that a nucleon moves inside a certain potential well (which keeps it bound to the nucleus) independently from the other nucleons. This amounts to replacing a N-body problem (N particles interacting) by N single-body problems. This essential simplification of the problem is the cornerstone of mean field theories. These are also widely used in atomic physics, where electrons move in a mean field due to the central nucleus and the electron cloud itself.

The independent particle model and mean field theories (we shall see that there exist several variants) have a great success in describing the properties o the nucleus starting from an effective interaction or an effective potential, thus are a basic part of atomic nucleus theory. One should also notice that they are modular enough, in that it is quite easy to extend the model to introduce effects such as nuclear pairing, or collective motions of the nucleon like rotation, or vibration, adding the corresponding energy terms in the formalism. This implies that in many representation the mean field is only a starting point for a more complete description which introduces correlations reproducing properties like collective excitations and nucleon transfer.[5][6]

10.3.2   Nuclear potential and effective interaction

A large part of the practical difficulties met in mean field theories is the definition (or calculation) of the potential of the mean field itself. One can very roughly distinguish between two approaches :

- The **phenomenological** approach is a parameterization of the nuclear potential by an appropriate mathematical function. Historically, this procedure was applied with the greatest success by Sven Gösta Nilsson, who used as a potential a (deformed) harmonic oscillator potential. The most recent parameterizations are based on more realistic functions, which account more accurately for scattering experiments, for example. In particular the form known as the Woods Saxon potential can be mentioned.

- The **self-consistent** or Hartree–Fock approach aims to deduce mathematically the nuclear potential from an effective nucleon–nucleon interaction. This technique implies a resolution of the Schrödinger equation in an iterative fashion, starting from an ansatz wavefunction and improving it variationally, since the potential depends there upon the wavefunctions to be determined. The latter are written as Slater determinants.

In the case of the Hartree–Fock approaches, the trouble is not to find the mathematical function which describes best the nuclear potential, but that which describes best the nucleon–nucleon interaction. Indeed, in contrast with atomic physics where the interaction is known (it is the Coulomb interaction), the nucleon–nucleon interaction within the nucleus is not known analytically.

There are two main reasons for this fact. First, the strong interaction acts essentially among the quarks forming the nucleons. The nucleon–nucleon interaction in vacuum is a mere *consequence* of the quark–quark interaction. While the latter is well understood in the framework of the Standard Model at high energies, it is much more complicated in low energies due to color confinement and asymptotic freedom. Thus there is yet no fundamental theory allowing one to deduce the nucleon–nucleon interaction from the quark–quark interaction. Furthermore, even if this problem were solved, there would remain a large difference between the ideal (and conceptually simpler) case of two nucleons interacting in vacuum, and that of these nucleons interacting in the nuclear matter. To go further, it was necessary to invent the concept of effective interaction. The latter is basically a mathematical function with several arbitrary parameters, which are adjusted to agree with experimental data.

Most modern interaction are zero-range so they act only when the two nucleons are in contact, as introduced by Tony Skyrme.[7]

10.3.3   The self-consistent approaches of the Hartree–Fock type

In the Hartree–Fock approach of the n-body problem, the starting point is a Hamiltonian containing n kinetic energy terms, and potential terms. As mentioned before, one of the mean field theory hypotheses is that only the two-body interaction is to be taken into account. The potential term of the Hamiltonian represents all possible two-body interactions in the set of n fermions. It is the first hypothesis.

The second step consists in assuming that the wavefunction of the system can be written as a Slater determinant of one-particle spin-orbitals. This statement is the mathematical translation of the independent-particle model. This is the second hypothesis.

There remains now to determine the components of this Slater determinant, that is, the individual wavefunctions of the nucleons. To this end, it is assumed that the total wavefunction (the Slater determinant) is such that the energy is minimum. This is the third hypothesis.

Technically, it means that one must compute the mean value of the (known) two-body Hamiltonian on the (unknown) Slater determinant, and impose that its mathematical variation vanishes. This leads to a set of equations where the unknowns are the individual wavefunctions: the Hartree–Fock equations. Solving these equations gives the wavefunctions and individual energy levels of nucleons, and so the total energy of the nucleus and its wavefunction.

This short account of the Hartree–Fock method explains why it is called also the variational approach. At the beginning of the calculation, the total energy is a "function of the individual wavefunctions" (a so-called functional), and everything is then made in order to optimize the choice of these wavefunctions so that the functional has a minimum – hopefully absolute, and not only local. To be more precise, there should be mentioned that the energy is a functional of the density, defined as the sum of the individual squared wavefunctions. Let us note also that the Hartree–Fock method is also used in atomic physics and condensed matter physics as Density Functional Theory, DFT.

The process of solving the Hartree–Fock equations can only be iterative, since these are in fact a Schrödinger equation in which the potential depends on the density, that is, precisely on the wavefunctions to be determined. Practically, the algorithm is started with a set of individual grossly reasonable wavefunctions (in general the eigenfunctions of a harmonic oscillator). These allow to compute the density, and therefrom the Hartree–Fock potential. Once this done, the Schrödinger equation is solved anew, and so on. The calculation stops – convergence is reached – when the difference among wavefunctions, or energy levels, for two successive iterations is less than a fixed value. Then the mean field potential is completely determined, and the Hartree–Fock equations become standard Schrödinger equations. The corresponding Hamiltonian is then called the Hartree–Fock Hamiltonian.

10.3.4   The relativistic mean field approaches

Born first in the 1970s with the works of D. Walecka on quantum hadrodynamics, the relativistic models of the nucleus were sharpened up towards the end of the 1980s by P. Ring and coworkers. The starting point of these approaches is the relativistic quantum field theory. In this context, the nucleon interactions occur via the exchange of virtual particles called mesons. The idea is, in a first step, to build a Lagrangian containing these interaction terms. Second, by an application of the least action principle, one gets a set of equations of motion. The real particles (here the nucleons) obey the Dirac equation, whilst the virtual ones (here the mesons) obey the Klein–Gordon equations.

In view of the non-perturbative nature of strong interaction, and also in view of the fact that the exact potential form of this interaction between groups of nucleons is relatively badly known, the use of such an approach in the case of atomic nuclei requires drastic approximations. The main simplification consists in replacing in the equations all field terms (which are operators in the mathematical sense) by their mean value (which are functions). In this way, one gets a system of coupled integro-differential equations, which can be solved numerically, if not analytically.

10.3.5   The interacting boson model

The interacting boson model (IBM) is a model in nuclear physics in which nucleons are represented as pairs, each of them acting as a boson particle, with integral spin of 0, 2 or 4. There are several branches of this model - in one of them (IBM-1) one can group all types of nucleons in pairs, in others (for instance - IBM-2) one considers protons and neutrons in pairs separately. See also: Interacting boson model.

10.3.6   Spontaneous breaking of symmetry in nuclear physics

One of the focal points of all physics is symmetry. The nucleon–nucleon interaction and all effective interactions used in practice have certain symmetries. They are invariant by translation (changing the frame of reference so that directions are not altered), by rotation (turning the frame of reference around some axis), or parity (changing the sense of axes) in the sense that the interaction does not change under any of these operations. Nevertheless, in the Hartree–Fock approach, solutions which are not invariant under such a symmetry can appear. One speaks then of spontaneous symmetry breaking.

Qualitatively, these spontaneous symmetry breakings can be explained in the following way : in the mean field theory, the nucleus is described as a set of independent particles. Most additional correlations among nucleons which do not enter the mean field are neglected. They can appear however by a breaking of the symmetry of the mean field Hamiltonian, which is only approximate. If the density used to start the iterations of the Hartree–Fock process breaks certain symmetries, the final Hartree–Fock Hamiltonian may break these symmetries, if it is advantageous to keep these broken from the point of view of the total energy.

It may also converge towards a symmetric solution. In any case, if the final solution breaks the symmetry, for example, the rotational symmetry, so that the nucleus appears not to be spherical, but elliptic, all configurations deduced from this deformed nucleus by a rotation are just as good solutions for the Hartree–Fock problem. The ground state of the nucleus is then *degenerate*.

A similar phenomenon happens with the nuclear pairing, which violates the conservation of the number of baryons (see below).

10.4   Extensions of the mean field theories

10.4.1   Nuclear pairing phenomenon

The most common extension to mean field theory is the nuclear pairing. Nuclei with an even number of nucleons are systematically more bound than those with an odd one. This implies that each nucleon binds with another one to form a pair, consequently the system cannot be described as independent particles subjected to a common a mean field. When the nucleus has an even number of protons and neutrons, each one of them finds a partner. To excite such a system, one must at least use such an energy as to break a pair. Conversely, in the case of odd number of protons or neutrons, there exists an unpaired nucleon, which needs less energy to be excited.

This phenomenon is closely analogous to that of Type 1 superconductivity in solid state physics. The first theoretical description of nuclear pairing was proposed at the end of the 1950s by Aage Bohr, Ben Mottelson, and David Pines (which contributed to the reception of the Nobel Prize in Physics in 1975 by Bohr and Mottelson).[8] It was close to the BCS theory of Bardeen, Cooper and Schrieffer, which accounts for metal superconductivity. Theoretically, the pairing phenomenon as described by the BCS theory combines with the mean field theory: nucleons are both subject to the mean field potential and to the pairing interaction.

Hartree–Fock–Bogolyubov (HFB) approach is a more sophisticated approach,[9] that enable to consider the pairing and mean field interactions consistently on equal footing, and is now the de facto standard in the mean field treatment of nuclear systems.

10.4.2 Symmetry Restoration

Peculiarity of mean field methods is the calculation of nuclear property by explicit symmetry breaking. The calculation of the mean field with self-consistent methods (e.g. Hartree-Fock), breaks rotational symmetry, and the calculation of pairing property breaks particle-number.

Several techniques for symmetry restoration by projecting on good quantum numbers have been developed.[10]

10.4.3 Particle Vibrations Coupling

Mean field methods (eventually considering symmetry restoration) are a good approximation for the ground state of the system, even postulating a system of independent particles. Higher-order corrections consider the fact that the particles interact together by the means of correlation. These correlations can be introduced taking into account the coupling of independent particle degrees of freedom, low-energy collective excitation of systems with even number of protons and neutrons.

In these way excited states can be reproduced by the means of Random Phase Approximation (RPA), and eventually consistently calculating also corrections to the ground state (e.g. by the means of Nuclear Field Theory [6]).

10.5   See also

Nuclear magnetic moment

10.6   References

[1] von Weizsäcker, C. F. (1935). "Zur Theorie der Kernmassen". *Zeitschrift für Physik* (in German) **96** (7–8): 431–458. Bibcode:1935ZPhy...96..431W. doi:10.1007/BF01337700.

[2] Moeller, P.; Myers, W. D.; Swiatecki, W. J.; Treiner, J. (3 Sep 1984). "Finite Range Droplet Model". *Conference: 7. international conference on atomic masses and fundamental constants (AMCO-7), Darmstadt-Seeheim, F.R. Germany.*

[3] Sorlin, O.; Porquet, M.-G. (2008). "Nuclear magic numbers: New features far from stability". *Progress in Particle and Nuclear Physics* **61** (2): 602–673. arXiv:0805.2561. Bibcode:2008PrPNP..61..602S. doi:10.1016/j.ppnp.2008.05.001.

[4] Brink, David; Broglia, Ricardo A. (2005). *Nuclear Superfluidity*. Cambridge University Press.

[5] Ring, P.; Schuck, P. (1980). *The nuclear many-body problem*. Springer Verlag. ISBN 3-540-21206-X.

[6] http://arxiv.org/abs/1504.05335

[7] http://www.sciencedirect.com/science/article/pii/0375947475903383

[8] Broglia, Ricardo A.; Zelevinsky, Vladimir (2013). *Fifty Years of Nuclear BCS: Pairing in Finite Systems*. World Scientific. ISBN 978-981-4412-48-3.

[9] http://www.fuw.edu.pl/~{}dobaczew/hfbtho16w/node2.html

[10] B. F. Bayman, Nucl. Phys. 15, 33 (1960)

See also: Nuclear physics § References

10.6.1 General audience

- James M. Cork ; *Radioactivité & physique nucléaire*, Dunod (1949).

10.6.2 Introductory texts

- Luc Valentin ; *Le monde subatomique - Des quarks aux centrales nucléaires*, Hermann (1986).
- Luc Valentin ; *Noyaux et particules - Modèles et symétries*, Hermann (1997).
- David Halliday ; *Introductory Nuclear Physics*, Wiley & Sons (1957).
- Kenneth Krane ; *Introductory Nuclear Physics*, Wiley & Sons (1987).
- Carlos Bertulani ; *Nuclear Physics in a Nutshell*, Princeton University Press (2007).

10.6.3 Fundamental texts

- Peter E. Hodgson; *Nuclear Reactions and Nuclear Structure*. Oxford University Press (1971).
- Irving Kaplan; *Nuclear physics*, the Addison-Wesley Series in Nuclear Science & Engineering, Addison-Wesley (1956). 2nd edition (1962).
- A. Bohr & B. Mottelson ; *Nuclear Structure*, 2 vol., Benjamin (1969–1975). Volume 1 : *Single Particle Motion* ; Volume 2 : *Nuclear Deformations*. Réédité par World Scientific Publishing Company (1998), ISBN 981-02-3197-0.
- P. Ring & P. Schuck; *The nuclear many-body problem*, Springer Verlag (1980), ISBN 3-540-21206-X
- A. de Shalit & H. Feshbach; *Theoretical Nuclear Physics*, 2 vol., John Wiley & Sons (1974). Volume 1: *Nuclear Structure*; Volume 2: *Nuclear Reactions*, ISBN 0-471-20385-8

10.7 External links

English

- (English) Institut de Physique Nucléaire (IPN), France
- (English) Facility for Antiproton and Ion Research (FAIR), Germany
- (English) Gesellschaft für Schwerionenforschung (GSI), Germany
- (English) Joint Institute for Nuclear Research (JINR), Russia
- (English) Argonne National Laboratory (ANL), USA

- (English) Riken, Japan

- (English) National Superconducting Cyclotron Laboratory, Michigan State University, USA

- (English) Facility for Rare Isotope Beams, Michigan State University, USA

French

- (French) Institut de Physique Nucléaire (IPN), France

- (French) Centre de Spectrométrie Nucléaire et de Spectrométrie de Masse (CSNSM), France

- (French) Service de Physique Nucléaire CEA/DAM, France

- (French) Institut National de Physique Nucléaire et de Physique des Particules (In2p3), France

- (French) Grand Accélérateur National d'Ions Lourds (GANIL), France

- (French) Commissariat à l'Energie Atomique (CEA), France

- (French) Centre Européen de Recherches Nucléaires, Suisse

-  **The LIVEChart of Nuclides - IAEA**

See also: Nuclear physics § External links

Chapter 11

List of particles

This is a list of the different types of particles found or believed to exist in the whole of the universe. For individual lists of the different particles, see the list below.

11.1 Elementary particles

Main article: Elementary particle

Elementary particles are particles with no measurable internal structure; that is, they are not composed of other particles. They are the fundamental objects of quantum field theory. Many families and sub-families of elementary particles exist. Elementary particles are classified according to their spin. Fermions have half-integer spin while bosons have integer spin. All the particles of the Standard Model have been experimentally observed, recently including the Higgs boson.[1][2]

11.1.1 Fermions

Main article: Fermion

Fermions are one of the two fundamental classes of particles, the other being bosons. Fermion particles are described by Fermi–Dirac statistics and have quantum numbers described by the Pauli exclusion principle. They include the quarks and leptons, as well as any composite particles consisting of an odd number of these, such as all baryons and many atoms and nuclei.

Fermions have half-integer spin; for all known elementary fermions this is $1/2$. All known fermions, except neutrinos, are also Dirac fermions; that is, each known fermion has its own distinct antiparticle. It is not known whether the neutrino is a Dirac fermion or a Majorana fermion.[3] Fermions are the basic building blocks of all matter. They are classified according to whether they interact via the color force or not. In the Standard Model, there are 12 types of elementary fermions: six quarks and six leptons.

Quarks

Main article: Quark

Quarks are the fundamental constituents of hadrons and interact via the strong interaction. Quarks are the only known carriers of fractional charge, but because they combine in groups of three (baryons) or in groups of two with antiquarks (mesons), only integer charge is observed in nature. Their respective antiparticles are the antiquarks, which are identical except for the fact that they carry the opposite electric charge (for example the up quark carries charge $+2/3$, while the up antiquark carries charge $-2/3$), color charge, and baryon number. There are six flavors of quarks; the three positively charged quarks are called "up-type quarks" and the three negatively charged quarks are called "down-type quarks".

Leptons

Main article: Leptons

Leptons do not interact via the strong interaction. Their respective antiparticles are the antileptons which are identical, except for the fact that they carry the opposite electric charge and lepton number. The antiparticle of an electron is an antielectron, which is nearly always called a "positron" for historical reasons. There are six leptons in total; the three charged leptons are called "electron-like leptons", while the neutral leptons are called "neutrinos". Neutrinos are known to oscillate, so that neutrinos of definite flavor do not have definite mass, rather they exist in a superposition of mass eigenstates. The hypothetical heavy right-handed neutrino, called a "sterile neutrino", has been left off the list.

11.1.2 Bosons

Main article: Boson

Bosons are one of the two fundamental classes of particles, the other being fermions. Bosons are characterized by Bose–Einstein statistics and all have integer spins. Bosons may be either elementary, like photons and gluons, or composite, like mesons.

The fundamental forces of nature are mediated by gauge bosons, and mass is believed to be created by the Higgs field. According to the Standard Model the elementary bosons are:

The graviton is added to the list although it is not predicted by the Standard Model, but by other theories in the framework of quantum field theory. Furthermore, gravity is non-renormalizable. There are a total of eight independent gluons. The Higgs boson is postulated by the electroweak theory primarily to explain the origin of particle masses. In a process known as the "Higgs mechanism", the Higgs boson and the other gauge bosons in the Standard Model acquire mass via spontaneous symmetry breaking of the SU(2) gauge symmetry. The Minimal Supersymmetric Standard Model (MSSM) predicts several Higgs bosons. A new particle expected to be the Higgs boson was observed at the CERN/LHC on March 14, 2013, around the energy of 126.5GeV with an accuracy of close to five sigma (99.9999%, which is accepted as definitive). The Higgs mechanism giving mass to other particles has not been observed yet.

11.1.3 Hypothetical particles

Supersymmetric theories predict the existence of more particles, none of which have been confirmed experimentally as of 2014:

Note: just as the photon, Z boson and $W^{\pm}$ bosons are superpositions of the B^0, W^0, W^1, and W^2 fields – the photino, zino, and wino$^{\pm}$ are superpositions of the bino^0, wino^0, wino^1, and wino^2 by definition.

No matter if one uses the original gauginos or this superpositions as a basis, the only predicted physical particles are neutralinos and charginos as a superposition of them together with the Higgsinos.

Other theories predict the existence of additional bosons:

Mirror particles are predicted by theories that restore parity symmetry.

"Magnetic monopole" is a generic name for particles with non-zero magnetic charge. They are predicted by some GUTs.

"Tachyon" is a generic name for hypothetical particles that travel faster than the speed of light and have an imaginary rest mass.

Preons were suggested as subparticles of quarks and leptons, but modern collider experiments have all but ruled out their existence.

Kaluza–Klein towers of particles are predicted by some models of extra dimensions. The extra-dimensional momentum is manifested as extra mass in four-dimensional spacetime.

11.2   Composite particles

11.2.1   Hadrons

Main article: Hadron

Hadrons are defined as strongly interacting composite particles. Hadrons are either:

- Composite fermions, in which case they are called baryons.

- Composite bosons, in which case they are called mesons.

Quark models, first proposed in 1964 independently by Murray Gell-Mann and George Zweig (who called quarks "aces"), describe the known hadrons as composed of valence quarks and/or antiquarks, tightly bound by the color force, which is mediated by gluons. A "sea" of virtual quark-antiquark pairs is also present in each hadron.

Baryons

See also: List of baryons

Ordinary baryons (composite fermions) contain three valence quarks or three valence antiquarks each.

- Nucleons are the fermionic constituents of normal atomic nuclei:
  - Protons, composed of two up and one down quark (uud)
  - Neutrons, composed of two down and one up quark (ddu)
- Hyperons, such as the Λ, Σ, Ξ, and Ω particles, which contain one or more strange quarks, are short-lived and heavier than nucleons. Although not normally present in atomic nuclei, they can appear in short-lived hypernuclei.
- A number of charmed and bottom baryons have also been observed.

Some hints at the existence of exotic baryons have been found recently; however, negative results have also been reported. Their existence is uncertain.

- Pentaquarks consist of four valence quarks and one valence antiquark.

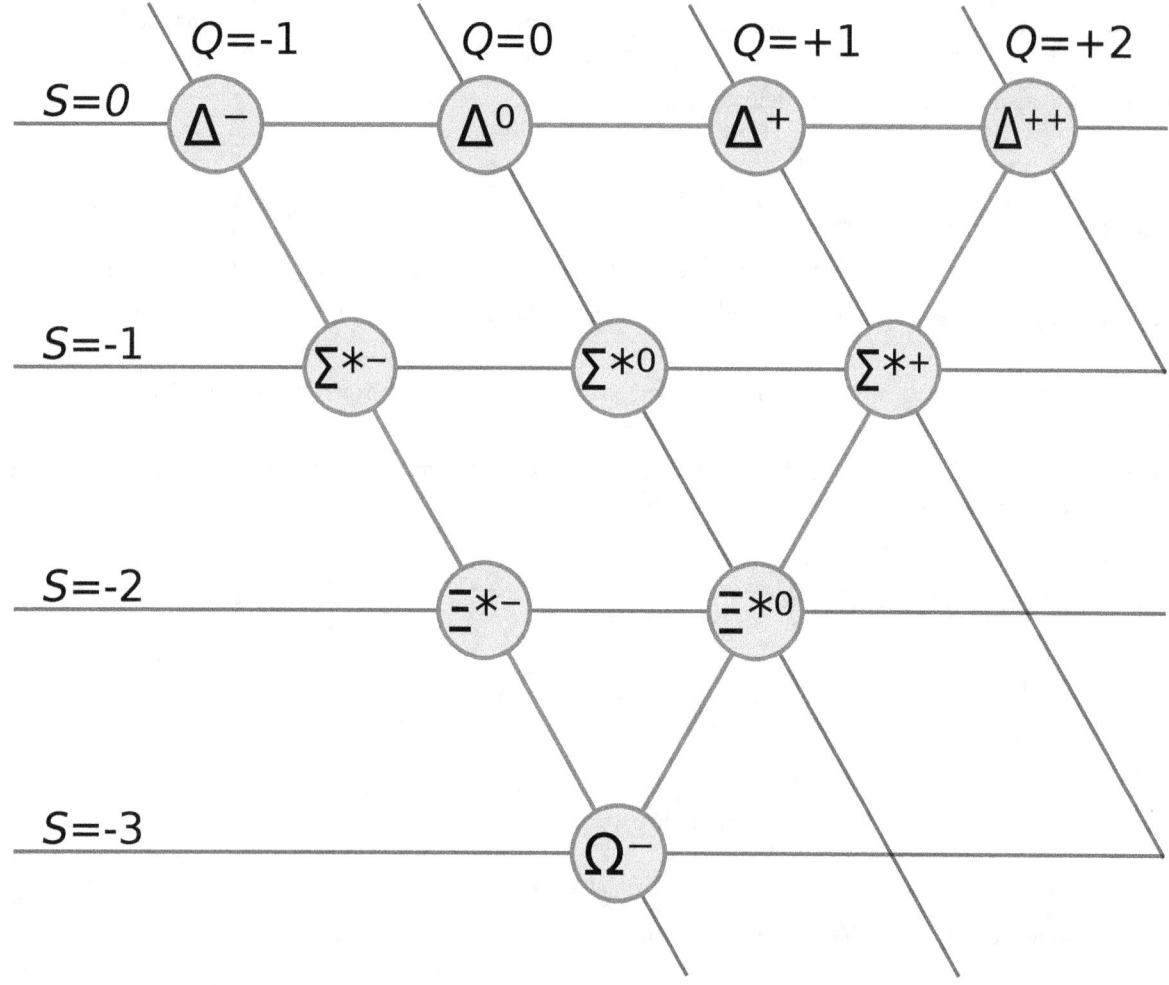

A combination of three u, d or s-quarks with a total spin of $^3\!/_2$ form the so-called "baryon decuplet".

Mesons

See also: List of mesons

Ordinary mesons are made up of a valence quark and a valence antiquark. Because mesons have spin of 0 or 1 and are not themselves elementary particles, they are "composite" bosons. Examples of mesons include the pion, kaon, and the J/ψ. In quantum hydrodynamic models, mesons mediate the residual strong force between nucleons.

At one time or another, positive signatures have been reported for all of the following exotic mesons but their existences have yet to be confirmed.

- A tetraquark consists of two valence quarks and two valence antiquarks;

- A glueball is a bound state of gluons with no valence quarks;

- Hybrid mesons consist of one or more valence quark-antiquark pairs and one or more real gluons.

11.2.2 Atomic nuclei

Atomic nuclei consist of protons and neutrons. Each type of nucleus contains a specific number of protons and a specific

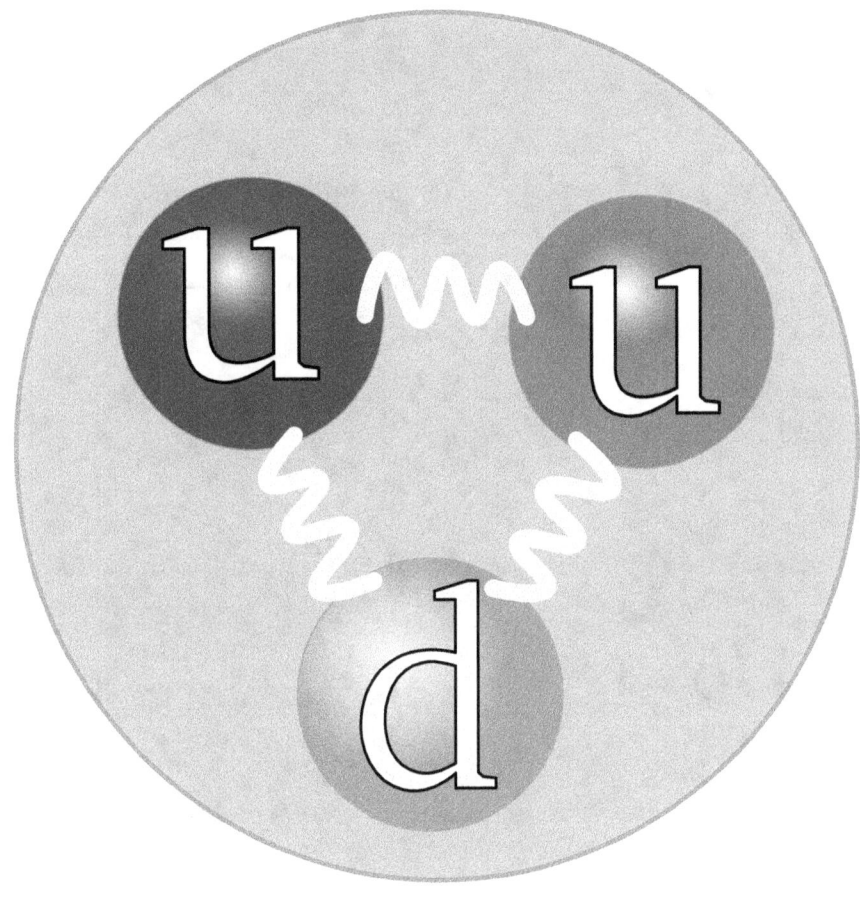

Proton quark structure: 2 up quarks and 1 down quark. The gluon tubes or flux tubes are now known to be Y shaped.

number of neutrons, and is called a "nuclide" or "isotope". Nuclear reactions can change one nuclide into another. See table of nuclides for a complete list of isotopes.

11.2.3 Atoms

Atoms are the smallest neutral particles into which matter can be divided by chemical reactions. An atom consists of a small, heavy nucleus surrounded by a relatively large, light cloud of electrons. Each type of atom corresponds to a specific chemical element. To date, 118 elements have been discovered, while only the elements 1-112,114, and 116 have received official names.

The atomic nucleus consists of protons and neutrons. Protons and neutrons are, in turn, made of quarks.

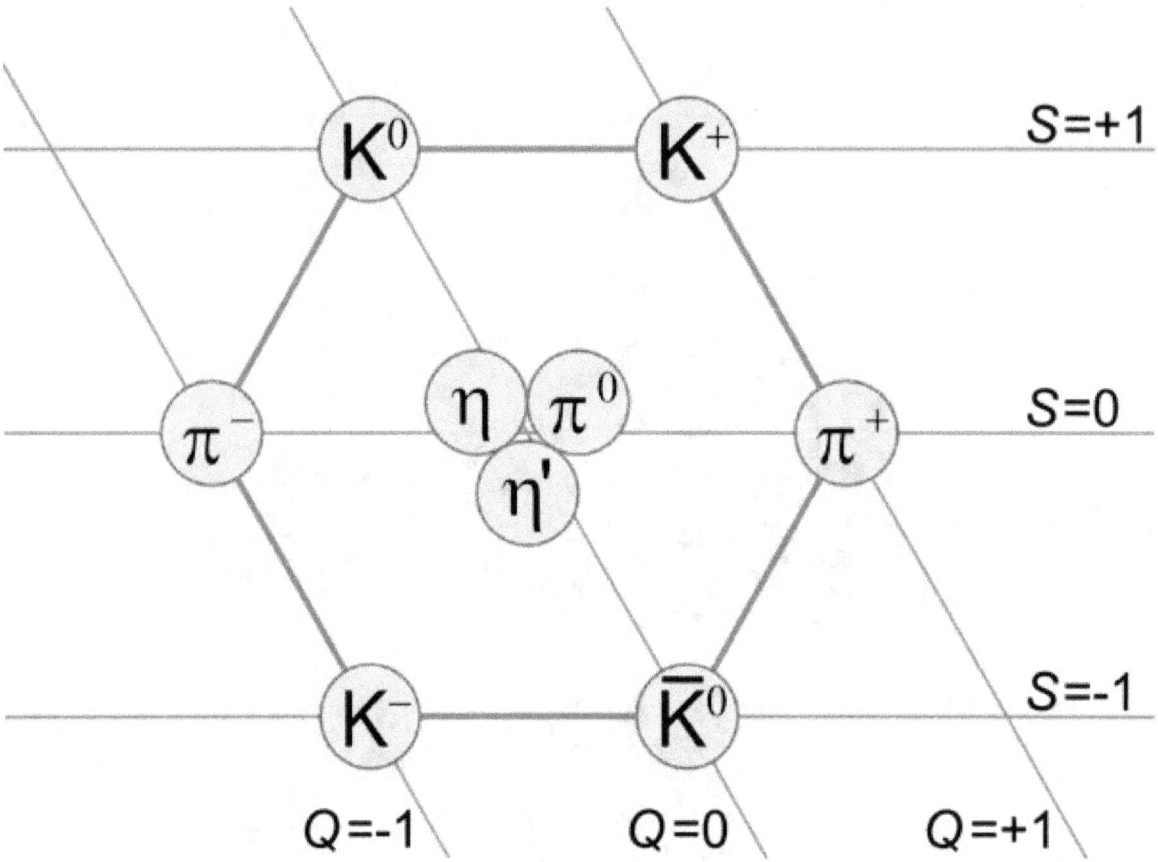

Mesons of spin 0 form a nonet

11.2.4 Molecules

Molecules are the smallest particles into which a non-elemental substance can be divided while maintaining the physical properties of the substance. Each type of molecule corresponds to a specific chemical compound. Molecules are a composite of two or more atoms. See list of compounds for a list of molecules.

11.3 Condensed matter

The field equations of condensed matter physics are remarkably similar to those of high energy particle physics. As a result, much of the theory of particle physics applies to condensed matter physics as well; in particular, there are a selection of field excitations, called quasi-particles, that can be created and explored. These include:

- Phonons are vibrational modes in a crystal lattice.

- Excitons are bound states of an electron and a hole.

- Plasmons are coherent excitations of a plasma.

- Polaritons are mixtures of photons with other quasi-particles.

- Polarons are moving, charged (quasi-) particles that are surrounded by ions in a material.

- Magnons are coherent excitations of electron spins in a material.

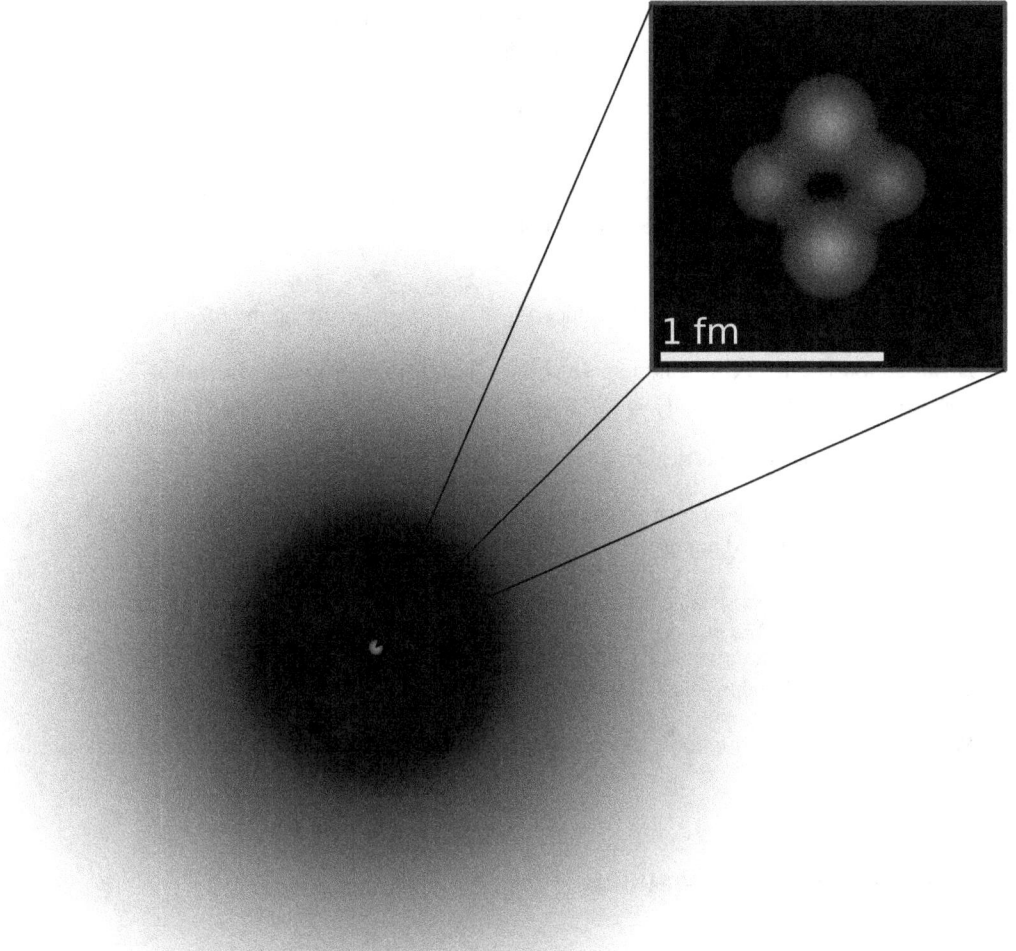

1 Å = 100,000 fm

A semi-accurate depiction of the helium atom. In the nucleus, the protons are in red and neutrons are in purple. In reality, the nucleus is also spherically symmetrical.

11.4   Other

- An anyon is a generalization of fermion and boson in two-dimensional systems like sheets of graphene that obeys braid statistics.

- A plekton is a theoretical kind of particle discussed as a generalization of the braid statistics of the anyon to dimension > 2.

- A WIMP (weakly interacting massive particle) is any one of a number of particles that might explain dark matter (such as the neutralino or the axion).

- The pomeron, used to explain the elastic scattering of hadrons and the location of Regge poles in Regge theory.

- The skyrmion, a topological solution of the pion field, used to model the low-energy properties of the nucleon, such as the axial vector current coupling and the mass.

- A genon is a particle existing in a closed timelike world line where spacetime is curled as in a Frank Tipler or Ronald Mallett time machine.

- A goldstone boson is a massless excitation of a field that has been spontaneously broken. The pions are quasi-goldstone bosons (quasi- because they are not exactly massless) of the broken chiral isospin symmetry of quantum chromodynamics.

- A goldstino is a goldstone fermion produced by the spontaneous breaking of supersymmetry.

- An instanton is a field configuration which is a local minimum of the Euclidean action. Instantons are used in nonperturbative calculations of tunneling rates.

- A dyon is a hypothetical particle with both electric and magnetic charges.

- A geon is an electromagnetic or gravitational wave which is held together in a confined region by the gravitational attraction of its own field energy.

- An inflaton is the generic name for an unidentified scalar particle responsible for the cosmic inflation.

- A spurion is the name given to a "particle" inserted mathematically into an isospin-violating decay in order to analyze it as though it conserved isospin.

- What is called "true muonium", a bound state of a muon and an antimuon, is a theoretical exotic atom which has never been observed.

11.5 Classification by speed

- A tardyon or bradyon travels slower than light and has a non-zero rest mass.

- A luxon travels at the speed of light and has no rest mass.

- A tachyon (mentioned above) is a hypothetical particle that travels faster than the speed of light and has an imaginary rest mass.

11.6 See also

- Acceleron

- List of baryons

- List of compounds for a list of molecules.

- List of fictional elements, materials, isotopes and atomic particles

- List of mesons

- Periodic table for an overview of atoms.

- Standard Model for the current theory of these particles.

- Table of nuclides

- Timeline of particle discoveries

11.7   References

[1] Observation of a new boson at a mass of 125 GeV with the CMS experiment at the LHC (2013). *arXiv:1207.7235*.

[2] Observation of a new particle in the search for the Standard Model Higgs boson with the ATLAS detector at the LHC (2012). *arXiv:1207.7214*.

[3] B. Kayser, *Two Questions About Neutrinos*, arXiv:1012.4469v1 [hep-ph] (2010).

[4] R. Maartens (2004). *Brane-World Gravity* (PDF). *Living Reviews in Relativity* **7**. p. 7. Also available in web format at http://www.livingreviews.org/lrr-2004-7.

- C. Amsler *et al.* (Particle Data Group) (2008). "Review of Particle Physics". *Physics Letters B* **667** (1–5): 1. Bibcode:2008PhLB..667....1P. doi:10.1016/j.physletb.2008.07.018. *(All information on this list, and more, can be found in the extensive, biannually-updated review by the Particle Data Group)*

Chapter 12

Binding energy

Binding energy is the energy required to disassemble a whole system into separate parts. A bound system typically has a lower potential energy than the sum of its constituent parts; this is what keeps the system together. Often this means that energy is released upon the creation of a bound state. This definition corresponds to a *positive* binding energy.

12.1   General idea

In general, binding energy represents the mechanical work that must be done against the forces which hold an object together, disassembling the object into component parts separated by sufficient distance that further separation requires negligible additional work.

At the atomic level the **atomic binding energy** of the atom derives from electromagnetic interaction and is the energy required to disassemble an atom into free electrons and a nucleus.[1] Electron binding energy is a measure of the energy required to free electrons from their atomic orbits. This is more commonly known as ionization energy.[2]

At the molecular level, bond energy and bond-dissociation energy are measures of the binding energy between the atoms in a chemical bond.

At the nuclear level, binding energy is also equal to the energy liberated when a nucleus is created from other nucleons or nuclei.[3][4] This nuclear binding energy (binding energy of nucleons into a nuclide) is derived from the nuclear force (residual strong interaction) and is the energy required to disassemble a nucleus into the same number of free, unbound neutrons and protons it is composed of, so that the nucleons are far/distant enough from each other so that the nuclear force can no longer cause the particles to interact.[5] Mass excess is a related concept which compares the mass number of a nucleus with its true measured mass.[6]

In astrophysics, the gravitational binding energy of a celestial body is the energy required to expand the material to infinity.

In bound systems, if the binding energy is removed from the system, it must be subtracted from the mass of the unbound system, simply because this energy *has* mass. Thus, if energy is removed (or emitted) from the system at the time it is bound, the loss of energy from the system will also result in the loss of the mass of the energy, from the system.[7] System mass is not conserved in this process because the system is "open" (i.e., is not an isolated system to mass or energy input or loss) during the binding process.

12.2   Mass-energy relation

Main articles: Mass–energy equivalence and Mass in special relativity

Classically a bound system is at a lower energy level than its unbound constituents, and its mass must be less than the total mass of its unbound constituents. For systems with low binding energies, this "lost" mass after binding may be fractionally small. For systems with high binding energies, however, the missing mass may be an easily measurable fraction. This missing mass may be lost during the process of binding as energy in the form of heat or light, with the removed energy corresponding to removed mass through Einstein's equation $E = mc^2$. Note that in the process of binding, the constituents of the system might enter higher energy states of the nucleus/atom/molecule, but these types of energy also have mass, and it is necessary that they be removed from the system before its mass may decrease. Once the system cools to normal temperatures and returns to ground states in terms of energy levels, there is less mass remaining in the system than there was when it first combined and was at high energy. In that case, the removed heat represents exactly the mass "deficit", and the heat itself retains the mass which was lost (from the point of view of the initial system). This mass appears in any other system which absorbs the heat and gains thermal energy.[8]

As an illustration, consider two objects attracting each other in space through their gravitational field. The attraction force accelerates the objects and they gain some speed toward each other converting the potential (gravity) energy into kinetic (movement) energy. When either the particles 1) pass through each other without interaction or 2) elastically repel during the collision, the gained kinetic energy (related to speed), starts to revert into potential form driving the collided particles apart. The decelerating particles will return to the initial distance and beyond into infinity or stop and repeat the collision (oscillation takes place). This shows that the system, which loses no energy, does not combine (bind) into a solid object, parts of which oscillate at short distances. Therefore, in order to bind the particles, the kinetic energy gained due to the attraction must be dissipated (by resistive force). Complex objects in collision ordinarily undergo inelastic collision, transforming some kinetic energy into internal energy (heat content, which is atomic movement), which is further radiated in the form of photons—the light and heat. Once the energy to escape the gravity is dissipated in the collision, the parts will oscillate at closer, possibly atomic, distance, thus looking like one solid object. This lost energy, necessary to overcome the potential barrier in order to separate the objects, is the binding energy. If this binding energy were retained in the system as heat, its mass would not decrease. However, binding energy lost from the system (as heat radiation) would itself have mass, and directly represents the "mass deficit" of the cold, bound system.

Closely analogous considerations apply in chemical and nuclear considerations. Exothermic chemical reactions in closed systems do not change mass, but become less massive once the heat of reaction is removed, though this mass change is much too small to measure with standard equipment. In nuclear reactions, however, the fraction of mass that may be removed as light or heat, i.e., binding energy, is often a much larger fraction of the system mass. It may thus be measured directly as a mass difference between rest masses of reactants and (cooled) products. This is because nuclear forces are comparatively stronger than the Coulombic forces associated with the interactions between electrons and protons, that generate heat in chemistry.

12.2.1 Mass change

Mass change (decrease) in bound systems, particularly atomic nuclei, has also been termed *mass defect*, *mass deficit*, or mass *packing fraction*.

The difference between the unbound system calculated mass and experimentally measured mass of nucleus (mass change) is denoted as Δm. It can be calculated as follows:

Mass change = (unbound system calculated mass) − (measured mass of system)

i.e., (sum of masses of protons and neutrons) − (measured mass of nucleus)

After nuclear reactions that result in an excited nucleus, the energy that must be radiated or otherwise removed as binding energy for a single nucleus to produce the unexcited state may be in any of several forms. This may be electromagnetic waves, such as gamma radiation, the kinetic energy of an ejected particle, such as an electron, in internal conversion decay, or partly as the rest mass of one or more emitted particles, such as the particles of beta decay. No mass deficit can in theory appear until this radiation or this energy has been emitted, and is no longer part of the system.

When nucleons bind together to form a nucleus, they must lose a small amount of mass, i.e., there is a change in mass, in order to stay bound. This mass change must be released as various types of photon or other particle energy as above, according to the relation $E = mc^2$. Thus, after binding energy has been removed, **binding energy = mass change × c^2**.

This energy is a measure of the forces that hold the nucleons together, and it represents energy that must be supplied again from the environment, if the nucleus were to be broken up into individual nucleons.

The energy given off during either nuclear fusion or nuclear fission is the difference between the binding energies of the "fuel", i.e., the initial nuclide(s), and the fission or fusion products. In practice, this energy may also be calculated from the substantial mass differences between the fuel and products, which uses previous measurement of the atomic masses of known nuclides, which always have the same mass for each species. This mass difference appears once evolved heat and radiation have been removed, which is a given requirement for measuring the (rest) masses of the (non-excited) nuclides involved in such calculations.

12.3 See also

- Chemical bond

- Electron binding energy

- Semi-empirical mass formula

- William Prout

- Virial mass

12.4 References

[1] "Nuclear Power Binding Energy". Retrieved 16 May 2015.

[2] IUPAC, *Compendium of Chemical Terminology*, 2nd ed. (the "Gold Book") (1997). Online corrected version: (2006–) "Ionization energy".

[3] *Britannica Online Encyclopaedia* - "nuclear binding energy". Accessed 8 September 2010. http://www.britannica.com/EBchecked/topic/65615/binding-energy

[4] Nuclear Engineering - "Binding Energy". Bill Garland, McMaster University. Accessed 8 September 2010. http://www.nuceng.ca/igna/binding_energy.htm

[5] *Atomic Alchemy: Nuclear Processes* - "Binding Energy". About. Accessed 7 September 2010. http://library.thinkquest.org/17940/texts/binding_energy/binding_energy.html

[6] Krane, K. S (1987). *Introductory Nuclear Physics*. John Wiley & Sons. ISBN 0-471-80553-X.

[7] *HyperPhysics* - "Nuclear Binding Energy". *C.R. Nave*, Georgia State University. Accessed 7 September 2010. http://hyperphysics.phy-astr.gsu.edu/hbase/nucene/nucbin.html

[8] E. F. Taylor and J. A. Wheeler, *Spacetime Physics*, W.H. Freeman and Co., NY. 1992. ISBN 0-7167-2327-1, see pp. 248-9 for discussion of mass remaining constant after detonation of nuclear bombs, until heat is allowed to escape.

12.5 External links

- Nuclear Binding energy

- Mass and Nuclide Stability

- Experimental atomic mass data compiled Nov. 2003

12.6   Text and image sources, contributors, and licenses

12.6.1   Text

- **Atomic nucleus** *Source:* https://en.wikipedia.org/wiki/Atomic_nucleus?oldid=682627820 *Contributors:* Kpjas, Andre Engels, XJaM, Merphant, Graft, Stevertigo, Patrick, JohnOwens, Michael Hardy, Tim Starling, Nixdorf, Bcrowell, Mcarling, Looxix~enwiki, Ellywa, Ahoerstemeier, Александър, Andres, Smack, Rednblu, The Anomebot, Topbanana, Cvaneg, Palefire, Gentgeen, Robbot, Sander123, Arkuat, Merovingian, Meelar, Wikibot, GarnetRChaney, Anthony, Giftlite, Graeme Bartlett, Christopher Parham, Awolf002, Mikez, Fastfission, Xerxes314, Bensaccount, Yath, Antandrus, Beland, OverlordQ, Icairns, Joyous!, JohnArmagh, Deglr6328, Trevor MacInnis, Mike Rosoft, Chris Howard, Perey, Guanabot, Igorivanov~enwiki, Vsmith, Gianluigi, Paul August, El C, Koenige, Madhu p, Shanes, Bookofjude, Mickeymousechen~enwiki, Foobaz, Obradovic Goran, Nsaa, Jumbuck, Disneyfreak96, Alansohn, Gintautasm, Neonumbers, Riana, Kfitzgib, Bart133, Velella, EvenT, H2g2bob, DV8 2XL, Mattbrundage, Kay Dekker, Flying fish, Stemonitis, Firsfron, AndriyK, Mandarax, Graham87, FreplySpang, Martinevos~enwiki, Rjwilmsi, Syndicate, Strait, Tangotango, Ttwaring, Lcolson, Nivix, OSt~enwiki, Maustrauser, Fresheneesz, Srleffler, Chobot, DVdm, Ahpook, Gwernol, Roboto de Ajvol, YurikBot, RobotE, Bambaiah, JWB, RussBot, Sillybilly, SpuriousQ, Hellbus, Okedem, Gaius Cornelius, CambridgeBayWeather, Wiki alf, FFLaguna, Dbfirs, Bota47, Supspirit, Tachyon01, Jess Riedel, Zzuuzz, Tsunaminoai, CWenger, Anclation~enwiki, Curpsbot-unicodify, Kungfuadam, RG2, AssistantX, SmackBot, FocalPoint, Incnis Mrsi, Bomac, Wogsland, Jrockley, Edgar181, Gilliam, Betacommand, Schmiteye, Chris the speller, Keegan, SchfiftyThree, DHN-bot~enwiki, Sbharris, Klacquement, Blake-, Itchjones, Dreadstar, DMacks, Daniel.Cardenas, Mion, Sadi Carnot, Bdushaw, Pilotguy, Kukini, Clicketyclack, Serein (renamed because of SUL), Kuru, Olin, Zarniwoot, NongBot~enwiki, Ekrub-ntyh, Funnybunny, MTSbot~enwiki, Cbuckley, Caiaffa, Dan Gluck, Kelvinaom, Joseph Solis in Australia, Lottamiata, Tubezone, Tawkerbot2, Flubeca, Sxim, Ale jrb, Scohoust, Shernren, Rowellcf, Engelmann15~enwiki, Kanags, MC10, Gogo Dodo, My Flatley, Christian75, DumbBOT, Bieeanda, Thijs!bot, Headbomb, Marek69, Tellyaddict, Wildthing61476, CTZMSC3, AntiVandalBot, Widefox, Quintote, TimVickers, Ilovescience, Gdo01, Gmarsden, JAnDbot, D99figge, Leuko, MER-C, Gfsheppard, .anacondabot, Bongwarrior, VoABot II, Rajb245, JamesBWatson, CalamusFortis, Mother.earth, Dirac66, Kopovoi, Dravick, Vssun, JaGa, Philg88, Hbent, Goodynotion, Akhil999in, MartinBot, Schmloof, Tiger-Smith, Roastytoast, J.delanoy, Classicalclarinet, Trusilver, Fnordius, Maurice Carbonaro, WarthogDemon, Lol nubs, Fylwind, Eshywiki, Tygrrr, TraceyR, Dylan bossart, VolkovBot, Indubitably, Stefan Kruithof, The Original Wildbear, Rei-bot, Anonymous Dissident, Piperh, Anna Lincoln, Shonenknifefan1, Venny85, Synthebot, Antixt, Enviroboy, PGWG, EmxBot, NEIL4737, SieBot, Sonicology, Scarian, Gerakibot, Caltas, Tiptoety, Onesspite, BenoniBot~enwiki, Afernand74, Cyfal, ClueBot, The Thing That Should Not Be, Jan1nad, Jekatz, Drmies, Boing! said Zebedee, Lainy8, CounterVandalismBot, Excirial, Alexbot, LordFoppington, SpikeToronto, Rhododendrites, Brews ohare, PhySusie, Tinymonty, Kakofonous, Versus22, InternetMeme, XLinkBot, Rangel lucy, Mr beeg lol, Addbot, Peyton.gaumer, Praseprase, Some jerk on the Internet, Leszek Jańczuk, Cst17, LinkFA-Bot, 5 albert square, CuteHappyBrute, Numbo3-bot, Tide rolls, Jan eissfeldt, Luckas-bot, Yobot, WikiDan61, 2D, TaBOT-zerem, Anypodetos, IW.HG, عالم محبوب, Podlif, Synchronism, AnomieBOT, Kingpin13, Flewis, Bluerasberry, Materialscientist, The High Fin Sperm Whale, Citation bot, Carlsotr, Quebec99, Xqbot, SouthH, Sionus, Capricorn42, Jeffrey Mall, RibotBOT, BSTemple, Shrikeangel, A. di M., Ironboy11, Kobewetnaps, Citation bot 1, Rylee118, Pinethicket, Tinton5, Games 101 wiki, RedBot, Minivip, White Shadows, FoxBot, Lionslayer, TheBFG, Dinamik-bot, Reaper Eternal, Specs112, Onel5969, RjwilmsiBot, Killaoftoast, DASHBot, EmausBot, Ornithikos, Mariov0288, The Pineapple, Cedar T., Thecheesykid, Hhhippo, JSquish, StringTheory11, Arbnos, Wayne Slam, Ocaasi, Kim cupcake, Sunshine4921, Mjbmrbot, Petrb, ClueBot NG, Magic Wizard, MelbourneStar, Fukushimayoshiho, Schunck, IOPhysics, Widr, Reify-tech, Helpful Pixie Bot, Downtowntrollin, Bibcode Bot, Lowercase sigmabot, AvocatoBot, Flying hippo705, Jordanf7, Gimp 11, Zedshort, Uopchem2511, Jburk711, DarafshBot, Akbask, JYBot, BrightStarSky, Dexbot, TwoTwoHello, Lugia2453, Bulba2036, Marekich, SassyLilNugget, Wlad2000, Mark viking, Siddhantsingh123, DennouNeko, DavidLeighEllis, Sladeb, Spyglasses, Y-S.Ko, KasparBot, Aless Val M and Anonymous: 441

- **Radioactive decay** *Source:* https://en.wikipedia.org/wiki/Radioactive_decay?oldid=682565072 *Contributors:* The Anome, Danny, Roadrunner, Mrwojo, Spiff~enwiki, Patrick, Ahoerstemeier, Andrewa, LittleDan, Kricke, Samw, Mxn, Smack, Hike395, HolIgor, Chuljin, Jitse Niesen, Audin, Furrykef, Populus, Omegatron, Topbanana, Pstudier, Finlay McWalter, PuzzletChung, Robbot, Romanm, Chancemill, Securiger, Merovingian, Pengo, Giftlite, Fudoreaper, Netoholic, Herbee, Everyking, Snowdog, Curps, Eequor, Jackol, Mmm~enwiki, Manuel Anastácio, Utcursch, Andycjp, LiDaobing, Antandrus, Beland, DragonflySixtyseven, Icairns, GeoGreg, Urhixidur, Syvanen, Olivier Debre, Deglr6328, Kate, Running, Mike Rosoft, Mormegil, Freakofnurture, Discospinster, Rydel, Rama, Vsmith, Mjpieters, Mani1, Night Gyr, Bender235, ESkog, Sunborn, Tompw, El C, J-Star, Lankiveil, Joanjoc~enwiki, Hayabusa future, RoyBoy, Orestes~enwiki, Grick, Bobo192, Stesmo, Smalljim, Indio~enwiki, Cohesion, Kjkolb, Nsaa, Storm Rider, Alansohn, Mr Adequate, AjAldous, Seans Potato Business, Ynhockey, Velella, Harej, RainbowOfLight, Dirac1933, Sciurinæ, Mikeo, DV8 2XL, Paraphelion, Zntrip, Ocollard, StradivariusTV, Duncan.france, Miss Madeline, CharlesC, Wdanwatts, Jacj, Qwertyus, Jclemens, Scuzzman, Martinevos~enwiki, Rjwilmsi, Jmcc150, Nneonneo, Bubba73, Watcharakorn, Lionelbrits, Ground Zero, Old Moonraker, RexNL, Kolbasz, Dalef, Fresheneesz, Guliolopez, Gwernol, Roboto de Ajvol, Wavelength, Phmer, Kymacpherson, RussBot, Jengelh, Shawn81, Kerowren, David Woodward, Gaius Cornelius, CambridgeBayWeather, Rsrikanth05, Bovineone, Tungsten, Grafen, Jaxl, Welsh, ONEder Boy, Ino5hiro, DJ John, Lomn, Scottfisher, DeadEyeArrow, Jeremy Visser, Ignitus, Wknight94, FF2010, Light current, Sefarkas, Closedmouth, Жованьб, Reyk, CharlesHBennett, CWenger, Fourohfour, Caco de vidro, Moomoomoo, Sbyrnes321, DVD R W, CIreland, Xtraeme, Eog1916, Itub, MacsBug, SmackBot, FocalPoint, Jclerman, Lcarsdata, Incnis Mrsi, KnowledgeOfSelf, Joonhon, Hydrogen Iodide, NoahWolfe, Jmulvey, Blue520, CMD Beaker, Jrockley, Yamaguchi先生, Gilliam, Carl.bunderson, TRosenbaum, Ati3414, Chris the speller, Bluebot, Kurykh, Agateller, Cadmium, MK8, Metacomet, Uthbrian, Reko, Sbharris, Rogermw, NYKevin, Can't sleep, clown will eat me, Ajaxkroon, Shalom Yechiel, Abyssal, V1adis1av, Ioscius, KaiserbBot, Rrburke, VMS Mosaic, Rsm99833, Addshore, Mrdempsey, Megamix, Flyguy649, Smooth O, Xyzzy n, Dreadstar, -Ozone-, Lcarscad, Cockneyite, Drphilharmonic, DMacks, Where, Bidabadi~enwiki, Cyberevil, Lambiam, SuperTycoon, Sanya, JoshuaZ, Accurizer, Minna Sora no Shita, IronGargoyle, 16@r, Ryulong, Peyre, Squirepants101, Dan Gluck, BranStark, Pegasus1138, CP\M, Freelance Intellectual, Fdp, Tawkerbot2, Chetvorno, Bstepp99, Conrad.Irwin, INkubusse, Xcentaur, RSido, Vyznev Xnebara, Nunquam Dormio, Solargenerator9.5, MarsRover, Leujohn, Smoove Z, Myasuda, J. Tyler, Island Dave, Quinnculver, Kanags, Gogo Dodo, HPaul, Mad-rick, Rracecarr, Skittleys, Christian75, FastLizard4, Gmoney650, The real avenger, Mikewax, Thijs!bot, Epbr123, Plmoknijb, Dougsim, Headbomb, Marek69, Deschreiber, Davidhorman, Meteoritekid, FourBlades, Stannered, Mentifisto, AntiVandalBot, Quintote, Jj137, Panu Petteri Höglund, Hanzoro5, Myanw, JAnDbot, Arch dude, Andonic, Xact, Snowynight, Acroterion, Geniac, Freedomlinux, Bongwarrior, VoABot II, AuburnPilot, Hillgentleman, JNW, Estonofunciona~enwiki, DMcanada, Klausok, Pixel ;-), Colinsweet, SparrowsWing, Indon, Animum, Dirac66, 28421u2232nfenfcenc, Loren-

erybody, Wiknerd, Dolphin51, Denisarona, ArepoEn, ClueBot, NickCT, Avenged Eightfold, Binksternet, Shardwing, Panoptik, The Thing That Should Not Be, VsBot, Arakunem, VQuakr, Rosuav, Uncle Milty, Canadianfelix, Otolemur crassicaudatus, Neverquick, DragonBot, Excirial, CohesionBot, Andy pyro, Bestunderblue, Lartoven, Jotterbot, PhySusie, Iohannes Animosus, Razorflame, BrandonLovesCake, Thingg, Aitias, Jonverve, Versus22, NJGW, Ginbot86, Nskrill, Vanished user 01, Jamesscottbrown, XLinkBot, Stick2k7, Oldnoah, Little Mountain 5, Avoided, Mifter, ZooFari, Hermanoere, Shoemaker's Holiday, Nuclear fusion man, Addbot, Bill344, Grotheconnor, Jacopo Werther, AVand, Some jerk on the Internet, Slimeyy, Ronhjones, Laurinavicius, Glane23, Favonian, Weekwhom, 5 albert square, Tide rolls, ScAvenger, David0811, KitemanSA, Johnsteinbeck2008, Mr. Soju, Luckas-bot, TheSuave, Yobot, WikiDan61, Andreasmperu, Fraggle81, Edoe, Nallimbot, KamikazeBot, Ayrton Prost, Azcolvin429, Eric-Wester, Tempodivalse, AnomieBOT, Andrewrp, Celsius100, Piano non troppo, AdjustShift, Ulric1313, Materialscientist, Phoenix of9, The High Fin Sperm Whale, Citation bot, Kalamkaar, Modesto Montoya, GB fan, ArthurBot, Cfwoodbury, Quebec99, LilHelpa, Xqbot, Tripodian, General3322, Addihockey10, Capricorn42, DSisyphBOT, NFD9001, Novonium, Hi878, GrouchoBot, Backpackadam, The Interior, Sternmusik, Sophus Bie, Shadowjams, Aaron2571, A.amitkumar, Bekus, StoneProphet, Hyperboreer, Zcoolz, FrescoBot, LucienBOT, KerryO77, EnglishAir, Gatlin86, Finalius, Citation bot 1, Pinethicket, PrincessofLlyr, Chatfecter, LinDrug, Camaxxy, BigDwiki, A8UDI, Footwarrior, Alex146, Jauhienij, IVAN3MAN, MusicNewz, Etincelles, Ipsoko, Comet Tuttle, Vrenator, Mr.98, AlexOVRLORD, Max Janu, ThinkEnemies, Sampathsris, Minimac, Hullernuc, Bhawani Gautam, NerdyScienceDude, Salvio giuliano, Slon02, Tacosrawesome, DASHBot, John of Reading, Adrenilyze, Dewritech, RA0808, Marcus Alan Young (1979), Masonrose, Your Lord and Master, K6ka, Hhhippo, Cjwinchester, JSquish, Josve05a, Eilishsholai, Skrapi28, Azuris, Quondum, Dasekely, IIIraute, Bushmillsmccallan, Dorothyzbornak, Wayne Slam, Δ, Sailsbystars, Carmichael, Orange Suede Sofa, Bomazi, Yaboyinc, Whoop whoop pull up, ClueBot NG, This lousy T-shirt, Slushee Xii, Widr, Shandirockwell, PrincessWortheverything, Sobieski Wanda, NuclearEnergy, Helpful Pixie Bot, Kid195631, Geo7777, Bibcode Bot, Krenair, MusikAnimal, Mark Arsten, Lazord00d, Dipankan001, Lesscoolroy, Seanisagod, Vkoves, Blaspie55, Wakeel99, Tyrael123, $uperbadd420, Achowat, BattyBot, Cylonsareboss, ChrisGualtieri, GoShow, Arcandam, Gdrg22, BrightStarSky, Dexbot, Fifty53, Kkk5000, TwoTwoHello, Sweetsourav97, Graphium, Reatlas, Joeinwiki, Andreynosatov, Epicgenius, Morg00, JohnKreike, Bob123456789123, Telma1203, B14709, Nigstomper, Pian0man263, Glaisher, Anrnusna, Agent Button, TerryLongbowRev, Baus Nguyen, 7Sidz, Mahusha, Monkbot, Internucleon, Vieque, CATOLOG1, Thenapster1426, Trackteur, Thedarkrogue, Sbuttars17, Chickenbuttman, AnimaLEquinoX, Jessiestill, ZahraBWP, TheMagikCow, Tbell91, Sdb5555, BilCrobyJr, KasparBot, Djennings20, Gutzio, TheSixthHorseman and Anonymous: 942

- **Nucleosynthesis** *Source:* https://en.wikipedia.org/wiki/Nucleosynthesis?oldid=682288599 *Contributors:* Vicki Rosenzweig, Mav, Roadrunner, Artsygeek, Andres, Samw, Cherkash, Epo~enwiki, Dcoetzee, Reddi, Stone, Stormie, Cmbant, Korath, Arkuat, Rursus, Xanzzibar, GreatWhiteNortherner, Giftlite, Harp, Herbee, Curps, Gzornenplatz, Sidar, Karol Langner, Tdent, D6, Perey, Pjacobi, Vsmith, Eric Forste, TaintedMustard, Voxadam, Oliphaunt, Benbest, Rjwilmsi, Strait, Goudzovski, Fogelmatrix, Chobot, Spacepotato, Sir48, Beanyk, BeastRHIT, Uber nemo, Light current, Modify, GrinBot~enwiki, Cmglee, Nekura, SmackBot, KnowledgeOfSelf, Chris the speller, Sbharris, Colonies Chris, Siffler~enwiki, Krich, Nakon, OhioFred, JorisvS, Iridescent, Zaphody3k, Kurtan~enwiki, Colonel Marksman, Van helsing, MrFizyx, Lokal Profil, Myasuda, Hga, James E B, Agony, Thijs!bot, Epbr123, Headbomb, Gierszep, Orionus, IanOsgood, Dosbears, Jingxin, Jessicapierce, Marhault, J.delanoy, Xarqi, Drake Dun, Rex07, DorganBot, Idioma-bot, Aucitypops, 28bytes, ABF, Szymanda, Claydonald, JhsBot, UnitedStatesian, BotKung, Samuelih, Michael Frind, Scarian, Paradoctor, Breakyunit, LeoBC, Auntof6, DragonBot, Taxa, Shinkolobwe, HexaChord, Addbot, DOI bot, Climbingfool, Loupeter, Yobot, Reindra, Azcolvin429, AnomieBOT, Piano non troppo, Materialscientist, Citation bot, LilHelpa, Xqbot, Tucsoncasey, Mnmngb, Thehelpfulbot, FrescoBot, Citation bot 1, Tom.Reding, Double sharp, Extra999, Nucleosynthesis, Kaiomai, Androstachys, Egjensen, Hovgiv, Virtual Loïc, RockMagnetist, Starbuster39, Terraflorin, Llightex, ClueBot NG, Law of Entropy, Helpful Pixie Bot, BG19bot, M Behnia, Zedshort, BattyBot, Khazar2, Martiantenor, Mogism, James floodhall, Vanamonde93, UnTrueOrUnSimplified, Linuxjava, Trackteur, WAFred, Tetra quark, Boazkat, Larry8000 and Anonymous: 97

- **Nuclear physics** *Source:* https://en.wikipedia.org/wiki/Nuclear_physics?oldid=682554564 *Contributors:* AxelBoldt, Trelvis, Mav, The Anome, AstroNomer~enwiki, Andre Engels, Deb, SimonP, Peterlin~enwiki, Heron, Sylmobile, Ixfd64, Looxix~enwiki, Andres, Rob Hooft, Smack, Hashar, Tantalate, Zoicon5, Grendelkhan, Fibonacci, Raul654, Pstudier, Finlay McWalter, Robbot, Vespristiano, Securiger, Ojigiri~enwiki, Xanzzibar, Giftlite, Harp, Monedula, Spatch, Antandrus, Karol Langner, APH, Icairns, Karl-Henner, ErikNY, Iantresman, Tsemii, Edsanville, C14, Mike Rosoft, Jkl, Vsmith, El C, John Vandenberg, Maurreen, Nk, BW52, Obradovic Goran, MPerel, Nsaa, Ranveig, Jumbuck, Alansohn, Stevegiacomelli, Yhr, Riana, AjAldous, Malo, Snowolf, Velella, DV8 2XL, Zereshk, Blaze Labs Research, Linas, LOL, Polyparadigm, MONGO, Kmg90, Allen3, Dysepsion, Magister Mathematicae, BD2412, Emallove, Martinevos~enwiki, Rjwilmsi, Mayumashu, Koavf, Strait, Leeyc0, Lcolson, FayssalF, RexNL, Karelj, TheDJ, Srleffler, Smartsmith, Chobot, Celebere, DVdm, Hall Monitor, Whosasking, Roboto de Ajvol, TexasAndroid, Phmer, RussBot, Pigman, CambridgeBayWeather, Chaos, Rsrikanth05, David R. Ingham, Grafen, SCZenz, Dbfirs, Davidizer13, Junglecat, RG2, Pentasyllabic, DVD R W, Teo64x, SmackBot, Haymaker, Android 93, C.Fred, Jagged 85, Jab843, Kurykh, Persian Poet Gal, Bduke, Jfsamper, Mithaca, DHN-bot~enwiki, Sbharris, Colonies Chris, Lpgaffney, Onorem, Rrburke, Addshore, SundarBot, Theanphibian, Kkailas, Thejerm, JunCTionS, Kuru, Euchiasmus, Accurizer, Zarniwoot, IronGargoyle, Waggers, Courcelles, Wikipeedio, MrFizyx, Dgw, McVities, FlyingToaster, Scott.medling, Cydebot, Meighan, Gogo Dodo, HPaul, Corpx, Tpot2688, InfoCan, Pstanton, Headbomb, Pjvpjv, I already forgot, Austin Maxwell, Sidasta, AntiVandalBot, Seaphoto, Samprox, Wixim, Perikl~enwiki, Gillano, Yoosq, Fireice, Res2216firestar, D99figge, MER-C, CosineKitty, Magioladitis, Bongwarrior, VoABot II, MartinDK, UnknownThinker, Colinsweet, Vssun, DerHexer, BetBot~enwiki, Bissinger, Hairchrm, Science5, Tonbo0422, HEL, J.delanoy, Trusilver, C.A.T.S. CEO, Katalaveno, Ryan Postlethwaite, Pyrospirit, Pcfjr9, DAID, Treisijs, Grendlefuzz, TraceyR, Sheliak, Funandtrvl, Hugo999, VolkovBot, ABF, DSRH, Philip Trueman, Greatwalk, TXiKiBoT, The Original Wildbear, Steven J. Anderson, Leafyplant, Uknowme219, BotKung, Venny85, Madhero88, Syxxness, Insanity Incarnate, HiDrNick, AlleborgoBot, Logan, Starkrm, Kbrose, The Random Editor, SieBot, Tiddly Tom, BotMultichill, Gerakibot, RJaguar3, Bentogoa, Flyer22, Oda Mari, Momo san, Granf, Nuttycoconut, Iain99, Jack the Stripper, OKBot, Elliottcable, Moonside, Neo., Pinkadelica, StewartMH, Velvetron, ClueBot, Foxj, Zeropiel, ILLERT, Hbomb phd mom, Neverquick, Djr32, Excirial, TonyBallioni, Brews ohare, Jotterbot, Skylerb skyguy, Ember of Light, Chargers85, Xme, El bot de la dieta, Kikos, Vanished User 1004, Crazy Boris with a red beard, WikiHead, NellieBly, Mifter, PL290, Rangel lucy, Pmizzlemoneydizzle, Carguy85TA, Addbot, Himselflolol, Lohengriny, LX green, Fielddaysunday, CanadianLinuxUser, Leszek Jańczuk, Fluffernutter, Jpoelma13, WFPM, OliverTwisted, Protonk, Favonian, Mac8175, Tide rolls, Lightbot, Frehley, Legobot, Luckas-bot, Jcae6, Kvhs rules, 2D, Arow87, Fraggle81, Chatter, Cflm001, ArcticREPtilia, Lars Ruoff, محبوب, علام, AnomieBOT, Jim1138, Piano non troppo, Yangtairan, AdjustShift, Kingpin13, Materialscientist, Bci2, GB fan, ArthurBot, Htomfields, Gsmgm, Xqbot, 4twenty42o, Gfhfg, Ute in DC, Omnipaedista, SciberDoc, Ciceronibus, Alireza-aa, Ironboy11, KownMATS, EvusJeevus, XcoolmanX, D o z y, Þjóðólfr, Pinethicket, I dream of horses, Rushbugled13, Fat&Happy, Σ, Carolina cotton, FoxBot, Lotje, January, Dian-

Vanished user fijw983kjaslkekfhj45, Maschen, RockMagnetist, Stormymountain, Ζετα ζ, Whoop whoop pull up, Isocliff, ClueBot NG, Smtcha-hal, Snotbot, Tonypak, O.Koslowski, CharleyQuinton, Dsperlich, Theopolisme, ZakMarksbury, Helpful Pixie Bot, Bibcode Bot, BG19bot, Tirebiter78, AvocatoBot, Lukys~enwiki, Stapletongrey, Ownedroad9, Chip123456, ChrisGualtieri, Khazar2, Billyfesh399, Rhlozier, JYBot, Dexbot, Doom636, Rongended, Cerabot~enwiki, CuriousMind01, Cjean42, Jayanta mallick, Joeinwiki, Kowtje, JPaestpreornJeolhlna, Eye-snore, Euan Richard, Nigstomper, Particle physicist, Prokaryotes, Jernahthern, Ginsuloft, Dimension10, JNrgbKLM, Krabaey, 1codesterS, FelixRosch, Monkbot, Delbert7, BradNorton1979, Lathamboyle, Tetra quark, KasparBot, Buckbill10 and Anonymous: 357

- **Semi-empirical mass formula** *Source:* https://en.wikipedia.org/wiki/Semi-empirical_mass_formula?oldid=677619716 *Contributors:* Axel-Boldt, Charles Matthews, Dcoetzee, Giftlite, CyborgTosser, Rich Farmbrough, Vsmith, Xezbeth, Longhair, ABCD, Burn, Hdeasy, H2g2bob, Gene Nygaard, Christopher Thomas, Mike Peel, Hashproduct, Zapateria, Philten, YurikBot, Jimp, Krea, BirgitteSB, Nate1481, Sliggy, Smack-Bot, Dauto, Yurigerhard, Sbharris, Radagast83, Pwjb, Postscript07, NotoriousTF, Inquisitus, Dan Gluck, UberScienceNerd, Barticus88, TDF, Headbomb, Lovibond, Morngnstar, Dirac66, Mollwollfumble, Jim.henderson, Pamputt, YonaBot, Droog Andrey, ArdClose, Biggerj1, Niel.Bowerman, Tangoludwig, Addbot, Yobot, Quasar1826, AnomieBOT, ^musaz, Citation bot, Capricorn42, Dafirenze, Maggyero, Shash-wat986, Calmer Waters, Pbrower2a, Hullernuc, McSaks, Marsupilami (DE), GoingBatty, BredoteauU2, Jlemans89, Marechal Ney, Reify-tech, MerlIwBot, KKloepfer, BattyBot, Mogism, Aryaindia, Marekich, Joeinwiki, Bramlap92, Femmtom and Anonymous: 67

- **Nuclear shell model** *Source:* https://en.wikipedia.org/wiki/Nuclear_shell_model?oldid=682551997 *Contributors:* AxelBoldt, Andre Engels, Michael Hardy, Ahoerstemeier, Tantalate, Doradus, Zoicon5, Chuunen Baka, Donarreiskoffer, Gandalf61, Giftlite, Junkyardprince, Arivero, TheMile, Harley peters, Anthony Appleyard, Yhr, Hu, Dennis Bratland, Blaze Labs Research, Woohookitty, Benbest, Tabletop, Oldelpaso, Rjwilmsi, Rwirth, JWB, Jimp, Tetracube, SmackBot, Incnis Mrsi, Nipun jain, Sbharris, OrphanBot, DMacks, Dan Gluck, Jac16888, Cricket-girl, HPaul, Thijs!bot, Barticus88, WinBot, Jbom1, Bakken, Sikory, Adventurer, Vinograd19, OttoMäkelä, TraceyR, Hqb, Pamputt, SieBot, Cyfal, Naureenahsan, ClueBot, Johnuniq, Addbot, Damsgård, Lightbot, Traitor, Luckas-bot, Yobot, Robert Treat, Omnipaedista, Ciceronibus, HRoestBot, Carel.jonkhout, Double sharp, MagnInd, K6ka, Hhhippo, AManWithNoPlan, ClueBot NG, Lychung, Helpful Pixie Bot, Em-mie8992, BattyBot, Jimw338, Joeinwiki, Bramlap92, Joethepizza and Anonymous: 53

- **Nuclear structure** *Source:* https://en.wikipedia.org/wiki/Nuclear_structure?oldid=680355169 *Contributors:* Chuunen Baka, Michael Devore, OwenBlacker, ReelExterminator, Erkcan, Kebes, Siddhant, Jimp, Eleassar, Eigenlambda, SmackBot, Incnis Mrsi, Sbharris, OrphanBot, Eynar, Daniel.Cardenas, Trassiorf, Dr.K., Dan Gluck, Yellowstone6, Myasuda, Headbomb, Bm gub, Magioladitis, PhysicsIsh, TraceyR, Signalhead, VolkovBot, Piperh, Naureenahsan, ClueBot, Doctor Logic, DenverRedhead, Addbot, 13209hajfhd098, WFPM, Ben Ben, Luckas-bot, Yobot, AnomieBOT, FrescoBot, Tempmine, Thinking of England, Minivip, Gfoley4, Maschen, Eg-T2g, Schunck, Vowies, Bibcode Bot, BG19bot, Raghnar, Ragnarstroberg, BattyBot, Garuda0001, Joeinwiki, Trompedo, Inlehmann, Jbcfcj, Gear blank, Danbanani, Pt186quadrupole, The Quixotic Potato and Anonymous: 29

- **List of particles** *Source:* https://en.wikipedia.org/wiki/List_of_particles?oldid=682746251 *Contributors:* AxelBoldt, Danny, Rmhermen, Stevertigo, Bdesham, Ahoerstemeier, Stan Shebs, Docu, Salsa Shark, Nikai, Evercat, Schneelocke, Charles Matthews, Jitse Niesen, CBDunker-son, Bevo, Raul654, Donarreiskoffer, Robbot, Sanders muc, Merovingian, Pengo, Giftlite, Herbee, Xerxes314, Dratman, Jeremy Henty, Alen-sha, Bodhitha, Physicist, Hayne, Quadell, RetiredUser2, Mysidia, Icairns, Asbestos, D6, Urvabara, Discospinster, Rich Farmbrough, FT2, Qutezuce, ArnoldReinhold, Neko-chan, El C, Laurascudder, Susvolans, EmilJ, Physicistjedi, Minghong, Gbrandt, Eddideigel, Axl, Mac Davis, David Ko, Radical Mallard, RJFJR, Count Iblis, Dirac1933, TenOfAllTrades, LFaraone, Oleg Alexandrov, Linas, JarlaxleArtemis, Dun-can.france, GregorB, Cedrus-Libani, Karam.Anthony.K, Palica, Rjwilmsi, Zbxgscqf, JLM~enwiki, Strait, Ems57fcva, Krash, Dan Guan, Dan-nyWilde, Lmatt, Goudzovski, Chobot, YurikBot, Bambaiah, Vuvar1, Madkayaker, Hydrargyrum, Presscorr, Chaos, Salsb, Tavilis, SCZenz, Lexicon, TUSHANT JHA, Dna-webmaster, Tomvds, Poulpy, Cstmoore, TLSuda, NeilN, MacsBug, Tom Lougheed, McGeddon, Bazza 7, WookieInHeat, Derdeib, Yamaguchi先生, Betacommand, Bluebot, Master of Puppets, DHN-bot~enwiki, Raistuumum, Juancnuno, Kittybrew-ster, Acepectif, Ligulembot, TriTertButoxy, ArglebargleIV, Khazar, John, FrozenMan, JorisvS, 041744, Dr Greg, Slakr, Mets501, Scor-pion0422, Cbuckley, Iridescent, TwistOfCain, Happy-melon, JRSpriggs, Flickboy, Van helsing, Lithium6, Neelix, Rotiro, Cydebot, Quibik, Christian75, Omicronpersei8, Thijs!bot, Qwyrxian, TauLibrus, Headbomb, Inner Earth, 49, Guptasuneet, Scottmsg, WinBot, Elmoosecapi-tan, Tyco.skinner, AubreyEllenShomo, Arch dude, Johnman239, Mwarren us, TheEditrix2, CalamusFortis, MartinBot, Sadisticsuburbanite, Bissinger, Anaxial, CommonsDelinker, Maurice Carbonaro, Zojj, OliverHarris, Joshmt, Adanadhel, Lseixas, Graphite Elbow, VolkovBot, Jmrowland, Quilbert, Anonymous Dissident, Dstary, Escalona, JPMasseo, Figureskatingfan, Inx272, Meters, Antixt, Hamish a e fowler, God-dersUK, Bluetryst, SieBot, Ishvara7, WereSpielChequers, Audrius u, VovanA, Paolo.dL, RSStockdale, Anchor Link Bot, StewartMH, Ex-plicit, ClueBot, Unbuttered Parsnip, Nolimitownass, DragonBot, Atomic7732, TimothyRias, SkyLined, Addbot, DOI bot, Jojhutton, Favonian, LinkFA-Bot, OlEnglish, Teles, Legobot, Luckas-bot, Yobot, Dov Henis, Azcolvin429, AnomieBOT, Götz, Icalanise, Flewis, Materialscientist, OllieFury, Vuerqex, ArthurBot, Vulcan Hephaestus, Blennow, Reality006, Coretheapple, Jcimorra, RibotBOT, Ernsts, A. di M., Axelfoley12, Zosterops, FrescoBot, Paine Ellsworth, Citation bot 1, JIK1975, Tom.Reding, Diffequa, WikitanvirBot, Racerx11, 112358sam, Aegnor.erar, Hops Splurt, HESUPERMAN, Hhhippo, AvicBot, JSquish, StringTheory11, Waperkins, Bamyers99, Suslindisambiguator, L Kensington, Den-nisIsMe, RockMagnetist, ClueBot NG, Snotbot, Primergrey, Vio45lin, Widr, MsFionnuala, Oklahoma3477, Bibcode Bot, CityOfSilver, Cap'n G, BML0309, Dan653, Twocount, Penguinstorm300, Dexbot, LightandDark2000, Ohiggy, TwoTwoHello, Andyhowlett, Printersmoke, Orion 2013, ARUNEEK, Seino van Breugel, AspaasBekkelund, TheMagikCow, Vyom27, ParkersComments, Selva Ganapathy and Anonymous: 290

- **Binding energy** *Source:* https://en.wikipedia.org/wiki/Binding_energy?oldid=667349242 *Contributors:* Bryan Derksen, Patrick, Tim Starling, Kku, SebastianHelm, The Anomebot, Taxman, Robbot, Fredrik, Securiger, Auric, Harp, BenFrantzDale, Fastfission, LeYaYa, Jason Quinn, ConradPino, Creidieki, Pinnerup, Zowie, RJHall, Cap'n Refsmmat, Neilrieck, CDN99, Mike Schwartz, La goutte de pluie, Kjkolb, Larry V, Nsaa, Jjron, Jhd, Riana, EagleFalconn, H2g2bob, Zereshk, Capecodeph, Blaze Labs Research, Firsfron, Reinoutr, Woohookitty, Cruccone, Robert K S, Rjwilmsi, Wikibofh, Kazrak, FlaBot, Margosbot~enwiki, Fragglet, Gurch, Fresheneesz, DVdm, Dstrozzi, YurikBot, Wavelength, Spacepotato, JWB, Van der Hoorn, Wikimachine, Howcheng, Goffrie, Kkmurray, Covington, Petri Krohn, Benonemusic, CWenger, RG2, Phr en, Bibliomaniac15, SmackBot, Jrockley, RobotJcb, Fetofs, Croquant, Sbharris, Colonies Chris, Javalenok, OrphanBot, Voyajer, Argle-bargleIV, Sophia, Euchiasmus, Herr apa, Zarniwoot, Dan Gluck, Igoldste, JRSpriggs, Cydebot, Nick Y., Bvcrist, DumbBOT, Thijs!bot, Bar-ticus88, Headbomb, PJtP, EdJohnston, WhaleyTim, Canarris, Seaphoto, CPWinter, Jayron32, Larrybaxter, Kiliman, Bongwarrior, VoABot II, Luctuosa, Stratford15, Mollwollfumble, Su-no-G, Robin S, Natsirtguy, T.vanschaik, Panguanwen, Gombang, Rominandreu, Brian Pearson, BigHairRef, Potatoswatter, MetsFan76, BrianScanlan, Kamparius, Signalhead, VolkovBot, TXiKiBoT, A4bot, BotKung, YohanN7, Caltas, Jimmycleveland, Proton666, Dabears36, ClueBot, Boing! said Zebedee, Erebus Morgaine, Jamespitt, Iohannes Animosus, SoxBot III, Gnowor,

Addbot, DOI bot, WFPM, NjardarBot, Kyle1278, Tide rolls, Legobot, Luckas-bot, Yobot, AnomieBOT, Xqbot, GrouchoBot, Pepemonbu, Sumankumardutta, A. di M., Dave3457, Darkskynet, I dream of horses, Abductive, Tom.Reding, Achim1999, Gistmass, V.V.93, Engineer-FromVega, Chriss.2, Techhead7890, Ibbn, John Cline, Mast3rj3di, Quondum, Xrayburst1, ClueBot NG, CocuBot, Wikihelperhelper, Silvrous, Lmfaoolmfaa, Radoslaw.c, Makecat-bot, DSCrowned, Khuzema Ali, Jocelyndurrey and Anonymous: 141

12.6.2 Images

- **File:3D_anamation_of_the_Rutherford_atom.ogv***Source:*https://upload.wikimedia.org/wikipedia/commons/9/90/3D_anamation_of_ Rutherford_atom.ogv *License:* CC BY-SA 3.0 *Contributors:*

- AtomoRutherford.blend *Original artist:* Damek

- **File:Alfa_beta_gamma_radiation.svg** *Source:* https://upload.wikimedia.org/wikipedia/commons/d/d6/Alfa_beta_gamma_radiation.svg *License:* CC BY 2.5 *Contributors:* Traced from this PNG image. *Original artist:* User:Stannered

- **File:Alpha_Decay.svg** *Source:* https://upload.wikimedia.org/wikipedia/commons/7/79/Alpha_Decay.svg *License:* Public domain *Contributors:* This vector image was created with Inkscape. *Original artist:* Inductiveload

- **File:Ambox_important.svg** *Source:* https://upload.wikimedia.org/wikipedia/commons/b/b4/Ambox_important.svg *License:* Public domain *Contributors:* Own work, based off of Image:Ambox scales.svg *Original artist:* Dsmurat (talk · contribs)

- **File:Atom_diagram.png** *Source:* https://upload.wikimedia.org/wikipedia/commons/d/d8/Atom_diagram.png *License:* CC-BY-SA-3.0 *Contributors:* Transferred from en.wikipedia to Commons by Teratornis using CommonsHelper. *Original artist:* The original uploader was Fastfission at English Wikipedia

- **File:Baryon_decuplet.svg** *Source:* https://upload.wikimedia.org/wikipedia/commons/f/f6/Baryon_decuplet.svg *License:* Public domain *Contributors:* Own work (Original text: *self-made*) *Original artist:* Wierdw123 at English Wikipedia

- **File:Betheweizsaecker.jpg** *Source:* https://upload.wikimedia.org/wikipedia/commons/c/c0/Betheweizsaecker.jpg *License:* Public domain *Contributors:* Own work *Original artist:* Aluminium

- **File:Binding_energy_curve_-_common_isotopes.svg***Source:*https://upload.wikimedia.org/wikipedia/commons/5/53/Binding_energy_ -_common_isotopes.svg *License:* Public domain *Contributors:* ? *Original artist:* ?

- **File:CERN_LHC_Tunnel1.jpg** *Source:* https://upload.wikimedia.org/wikipedia/commons/f/fc/CERN_LHC_Tunnel1.jpg *License:* CC BY-SA 3.0 *Contributors:* Own work *Original artist:* Julian Herzog (website)

- **File:Commons-logo.svg** *Source:* https://upload.wikimedia.org/wikipedia/en/4/4a/Commons-logo.svg *License:* ? *Contributors:* ? *Original artist:* ?

- **File:Comparisons_of_other_mushroom_clouds.jpg***Source:*https://upload.wikimedia.org/wikipedia/commons/5/55/Comparisons_of_ mushroom_clouds.jpg*License:*Public domain*Contributors:*http://www.julg7.com/blog/wp-content/uploads/2009/02/tsar_bomba.jpg*Original artist:*http://julg7.com

- **File:Crookes_tube_xray_experiment.jpg** *Source:* https://upload.wikimedia.org/wikipedia/commons/1/10/Crookes_tube_xray_experiment. jpg *License:* Public domain *Contributors:* Downloaded 2007-12-23 from <a data-x-rel='nofollow' class='external text' href='http://books.google. com/books?id=whc4AAAAMAAJ,<span>,&,</span>,pg=PT5'>William J. Morton and Edwin W. Hammer (1896) *The X-ray, or Photography of the Invisible and its value in Surgery*, American Technical Book Co., New York, fig. 54</a> on Google Books *Original artist:* William J. Morton

- **File:Crystal_energy.svg** *Source:* https://upload.wikimedia.org/wikipedia/commons/1/14/Crystal_energy.svg *License:* LGPL *Contributors:* Own work conversion of Image:Crystal_128_energy.png *Original artist:* Dhatfield

- **File:DBP_1979_1020_Otto_Hahn_Kernspaltung.jpg** *Source:* https://upload.wikimedia.org/wikipedia/commons/8/83/DBP_1979_1020_ Otto_Hahn_Kernspaltung.jpg *License:* Public domain *Contributors:* scanned by NobbiP *Original artist:* Deutsche Bundespost

- **File:DecayRate_vs_Solar_Time.png***Source:*https://upload.wikimedia.org/wikipedia/commons/d/d3/DecayRate_vs_Solar_Time.png: Public domain *Contributors:* ? *Original artist:* ?

- **File:Elementary_particle_interactions_in_the_Standard_Model.png***Source:*https://upload.wikimedia.org/wikipedia/commons/a/a7/ particle_interactions_in_the_Standard_Model.png *License:* CC0 *Contributors:* Own work *Original artist:* Eric Drexler

- **File:Fission_chain_reaction.svg** *Source:* https://upload.wikimedia.org/wikipedia/commons/9/9a/Fission_chain_reaction.svg *License:* Public domain *Contributors:* ? *Original artist:* ?

- **File:Folder_Hexagonal_Icon.svg** *Source:* https://upload.wikimedia.org/wikipedia/en/4/48/Folder_Hexagonal_Icon.svg *License:* Cc-by-sa-3.0 *Contributors:* ? *Original artist:* ?

- **File:Gammaspektrum_Uranerz.jpg** *Source:* https://upload.wikimedia.org/wikipedia/commons/8/8c/Gammaspektrum_Uranerz.jpg *License:* CC BY-SA 3.0 *Contributors:* Own work *Original artist:* Wusel007

- **File:Halflife-sim.gif** *Source:* https://upload.wikimedia.org/wikipedia/commons/3/3f/Halflife-sim.gif *License:* Public domain *Contributors:* Own work *Original artist:* Sbyrnes321

- **File:He1523a.jpg** *Source:* https://upload.wikimedia.org/wikipedia/commons/5/5f/He1523a.jpg *License:* CC BY 4.0 *Contributors:* http:// www.solstation.com/x-objects/he1523.htm *Original artist:* ESO, European Southern Observatory

- **File:Helium_atom_QM.svg** *Source:* https://upload.wikimedia.org/wikipedia/commons/2/23/Helium_atom_QM.svg *License:* CC-BY-SA-3.0 *Contributors:* Own work *Original artist:* User:Yzmo

- **File:SolarSystemAbundances.png** *Source:* https://upload.wikimedia.org/wikipedia/commons/e/e6/SolarSystemAbundances.png *License:* CC BY-SA 3.0 *Contributors:* Transferred from en.wikipedia *Original artist:* Original uploader was 28bytes at en.wikipedia

- **File:Stagg_Field_reactor.jpg** *Source:* https://upload.wikimedia.org/wikipedia/commons/f/fe/Stagg_Field_reactor.jpg *License:* Public domain *Contributors:* http://narademo.umiacs.umd.edu/cgi-bin/isadg/viewobject.pl?object=95120 *Original artist:* Melvin A. Miller of the Argonne National Laboratory

- **File:Standard_Model_Feynman_Diagram_Vertices.png** *Source:* https://upload.wikimedia.org/wikipedia/commons/7/75/Standard_Model_Feynman_Diagram_Vertices.png *License:* CC BY-SA 3.0 *Contributors:* I made it in Adobe Illustrator *Original artist:* Garyzx

- **File:Standard_Model_of_Elementary_Particles.svg** *Source:* https://upload.wikimedia.org/wikipedia/commons/0/00/Standard_Model_of_Elementary_Particles.svg *License:* CC BY 3.0 *Contributors:* Own work by uploader, PBS NOVA [1], Fermilab, Office of Science, United States Department of Energy, Particle Data Group *Original artist:* MissMJ

- **File:Stdef2.png** *Source:* https://upload.wikimedia.org/wikipedia/commons/e/e0/Stdef2.png *License:* CC BY-SA 3.0 *Contributors:* Own work *Original artist:* Hullernuc

- **File:Stylised_Lithium_Atom.svg** *Source:* https://upload.wikimedia.org/wikipedia/commons/e/e1/Stylised_Lithium_Atom.svg *License:* CC-BY-SA-3.0 *Contributors:* based off of Image:Stylised Lithium Atom.png by Halfdan. *Original artist:* SVG by Indolences. Recoloring and ironing out some glitches done by Rainer Klute.

- **File:Symbol_book_class2.svg** *Source:* https://upload.wikimedia.org/wikipedia/commons/8/89/Symbol_book_class2.svg *License:* CC BY-SA 2.5 *Contributors:* Mad by Lokal_Profil by combining: *Original artist:* Lokal_Profil

- **File:Table_isotopes_en.svg** *Source:* https://upload.wikimedia.org/wikipedia/commons/c/c4/Table_isotopes_en.svg *License:* CC BY-SA 3.0 *Contributors:*

- Table_isotopes.svg *Original artist:* Table_isotopes.svg: Napy1kenobi

- **File:Text_document_with_red_question_mark.svg** *Source:* https://upload.wikimedia.org/wikipedia/commons/a/a4/Text_document_with_red_question_mark.svg *License:* Public domain *Contributors:* Created by bdesham with Inkscape; based upon Text-x-generic.svg from the Tango project. *Original artist:* Benjamin D. Esham (bdesham)

- **File:ThermalFissionYield.svg** *Source:* https://upload.wikimedia.org/wikipedia/commons/6/68/ThermalFissionYield.svg *License:* CC BY 3.0 *Contributors:* Transferred from en.wikipedia by SreeBot *Original artist:* JWB at en.wikipedia

- **File:UFission.gif** *Source:* https://upload.wikimedia.org/wikipedia/commons/8/86/UFission.gif *License:* CC BY-SA 3.0 *Contributors:* This file was created with Blender. *Original artist:* Anynobody

- **File:US_Atomic_Energy_Commission_logo.jpg** *Source:* https://upload.wikimedia.org/wikipedia/commons/7/7c/US_Atomic_Energy_logo.jpg *License:* Public domain *Contributors:* http://ma.mbe.doe.gov/me70/history/photos.htm *Original artist:* U.S. Department of Energy

- **File:Wikibooks-logo-en-noslogan.svg** *Source:* https://upload.wikimedia.org/wikipedia/commons/d/df/Wikibooks-logo-en-noslogan.svg *License:* CC BY-SA 3.0 *Contributors:* Own work *Original artist:* User:Bastique, User:Ramac et al.

- **File:Wikibooks-logo.svg** *Source:* https://upload.wikimedia.org/wikipedia/commons/f/fa/Wikibooks-logo.svg *License:* CC BY-SA 3.0 *Contributors:* Own work *Original artist:* User:Bastique, User:Ramac et al.

- **File:Wikinews-logo.svg** *Source:* https://upload.wikimedia.org/wikipedia/commons/2/24/Wikinews-logo.svg *License:* CC BY-SA 3.0 *Contributors:* This is a cropped version of Image:Wikinews-logo-en.png. *Original artist:* Vectorized by Simon 01:05, 2 August 2006 (UTC) Updated by Time3000 17 April 2007 to use official Wikinews colours and appear correctly on dark backgrounds. Originally uploaded by Simon.

- **File:Wikiquote-logo.svg** *Source:* https://upload.wikimedia.org/wikipedia/commons/f/fa/Wikiquote-logo.svg *License:* Public domain *Contributors:* ? *Original artist:* ?

- **File:Wikisource-logo.svg** *Source:* https://upload.wikimedia.org/wikipedia/commons/4/4c/Wikisource-logo.svg *License:* CC BY-SA 3.0 *Contributors:* Rei-artur *Original artist:* Nicholas Moreau

- **File:Wikiversity-logo-Snorky.svg** *Source:* https://upload.wikimedia.org/wikipedia/commons/1/1b/Wikiversity-logo-en.svg *License:* CC BY-SA 3.0 *Contributors:* Own work *Original artist:* Snorky

- **File:Wiktionary-logo-en.svg** *Source:* https://upload.wikimedia.org/wikipedia/commons/f/f8/Wiktionary-logo-en.svg *License:* Public domain *Contributors:* Vector version of Image:Wiktionary-logo-en.png. *Original artist:* Vectorized by Fvasconcellos (talk · contribs), based on original logo tossed together by Brion Vibber

12.6.3 Content license

www.ingramcontent.com/pod-product-compliance
Lightning Source LLC
Chambersburg PA
CBHW08081411180526
45168CB00006B/2439